Electrospinning - Material Technology of the Future

Edited by Tomasz Tański and Paweł Jarka

Published in London, United Kingdom

Electrospinning - Material Technology of the Future
http://dx.doi.org/10.5772/intechopen.97902
Edited by Tomasz Tański and Paweł Jarka

Contributors
Nilşen Sünter Eroğlu, Farai Dziike, Phylis Makurunje, Refilwe Matshitse, Alexandre A. Vetcher, Svetlana G. Karpova, Anna V. Bychkova, Anatoly A. Olkhov, Alexey L. Iordanskii, Azadeh Izadyari Aghmiuni, Arezzo Ghadi, Elmira Azmoun, Niloufar Kalantari, Iman Mohammadi, Hosein Hemati Kordmahaleh, Stanislav Petrik, Mayza Ibrahim, Cristian Javier Patiño Vidal, Marcelo Patiño Vidal, Carol Lopez de Dicastillo, Maria José Galotto, Cristina Muñoz-Shugulí

First published in London, United Kingdom, 2022 by IntechOpen
IntechOpen is the global imprint of INTECHOPEN LIMITED, registered in England and Wales, registration number: 11086078, 5 Princes Gate Court, London, SW7 2QJ, United Kingdom

British Library Cataloguing-in-Publication Data
A catalogue record for this book is available from the British Library

Additional hard and PDF copies can be obtained from orders@intechopen.com

Electrospinning - Material Technology of the Future
Edited by Tomasz Tański and Paweł Jarka
p. cm.
Print ISBN 978-1-80355-342-9
Online ISBN 978-1-80355-343-6
eBook (PDF) ISBN 978-1-80355-344-3

Meet the editors

Prof. Tomasz Tański is the head of the Department of Engineering Materials and Biomaterials, Silesian University of Technology, Poland, and a member of the Committee of Metallurgy of the Polish Academy of Sciences. He is also a specialist in non-ferrous alloys, composite and nanostructured materials, and structural engineering materials. He has authored or co-authored more than 400 scientific publications worldwide, including 15 monographs and books. He has won twenty national and international awards and honors. He is and/or was a supervisor or contractor for more than fifteen research and didactic projects in Poland and abroad. He is also a reviewer and promoter of numerous scientific papers, including eight doctoral research papers in the field of nanotechnology and materials.

Paweł Jarka obtained his Ph.D. from the Department of Engineering Materials and Biomaterials, Silesian University of Technology, Poland, where he is currently employed. The subject of his doctoral thesis was "The optoelectronic properties of active layers obtained by PVD methods in organic photovoltaic devices." His scientific interests concern functional materials. During his scientific activity, he has participated in many scientific international conferences. He is the co-author of several dozen scientific papers and five chapters in monographs and co-editor of one book. He also has one patent to his credit.

Contents

Preface

This book compiles up-to-date information on the innovative technique of producing nanostructures in the form of fibers, fibrous mats, and network structures. Continuous progress in many fields of science such as electronics, energy, medicine, and environmental protection is now driven by advances in the nanoscale, related primarily to the development of nanomaterials.

Among modern nanomaterials that enable the development of technologies and products, one-dimensional (1D) structures such as nanofibers and fibrous systems have an extremely wide range of applications. Hence, this book presents innovations in the production of 1D structures, such as electrospinning, as well as a whole range of innovative solutions obtained from dye to the use of one-dimensional nanostructures and their systems.

The first section of the book contains four chapters that discuss the use of electrospinning technology in the production of nanostructures of various types of materials, from commercially used polymers to synthetic waste biomass and plant extracts. In addition, some chapters broadly describe the parameters of the electrospinning process and analyze its impact on the structure and properties of the final nanostructures depending on application.

Chapter 1 describes the use of modern polymeric materials in the form of electrospun fibrous mats in the development of active food packaging as well as other applications in the food industry. Chapter 2 discusses the application of electrospinning technology in the production of fibrous mats used as substrates for the proliferation and differentiation of stem cells. Chapter 3 presents research on the production of structured nanofibers from natural sources, such as plant extracts. According to the chapter, the use of bio-additives in the form of natural herbal extracts in electrospun nanostructures may have extremely positive effects on biocompatibility and drug release possibilities. Chapter 4 presents a modern approach to the production and recycling of biomass waste streams. The chapter summarizes the use of the different types of biomass waste as an alternative source of polymeric materials in electrospinning technology.

The second section of the book describes the use of electrospun structures as "materials of the future" in areas with high development impact, including electronics and bioengineering. Chapter 5 contains literature reports and experimental data taken from the research on the production of fibrous materials used in controlled drug delivery systems. It examines the possibility of using electrospinning technology to improve material properties to achieve controlled release drug delivery systems. Chapter 6 presents an overview of novel material solutions in sensing applications with particular emphasis on nanofiber materials. According to the authors, a promising direction for the development of modern sensors is the use of functionalized nanofibers in hybrid fiber-optic/nanofiber structures, which presents the feasibility of realization of miniature sensors of biomedical and chemical values. The discussion in the chapter concerns the new detection principle for various chemical and biomedical applications.

Tomasz Tański and Paweł Jarka
Faculty of Mechanical Engineering,
Department of Engineering Materials and Biomaterials,
Silesian University of Technology,
Gliwice, Poland

Section 1

Technological Parameters of Electrospinning Technology in the Production of Functional Materials Nanostructures

Chapter 1

Active Electrospun Mats: A Promising Material for Active Food Packaging

Cristian Patiño Vidal, Cristina Muñoz-Shuguli,
Marcelo Patiño Vidal, María José Galotto
and Carol López de Dicastillo

Abstract

Nowadays, polymeric materials are widely used in the development of food packages. However, as food products with a greater safety and longer durability are required, packaging research area has been focused on the production of functional materials able to reach such further protection. The incorporation of natural and synthetics active compounds into the polymeric materials by traditional techniques has been the main used strategy, surging thus the research area of active food packaging. Furthermore, the latest science advances provide promising technologies for developing packaging materials, such as the electrospinning. This technique has allowed obtaining ultrathin electrospun mats based on micro- and/or nanofibers that have been proposed as novel active materials able to be applied as wrapper films, sachets and bags during the food packaging. In this chapter, the description of electrospinning, the effect of their principal parameters during the development of active food packaging materials as well as their current applications on different foodstuffs are presented.

Keywords: active compound, food package, food shelf life

1. Introduction

Packaging is one of the most useful tools used by the food industry to protect several foodstuffs against contamination and spoilage. Traditionally, materials such as plastic, glass, metals, paper and board have been used for developing food packages [1, 2]. However, the interest by polymeric materials have enormously increased because they exhibit several advantages, such as low-cost, low-weight and good mechanical, barrier and optical properties [3, 4]. Therefore, processed and non-processed foods are daily packaged into plastic materials in order to avoid their contamination by odors, dust and microorganisms, as well as their deterioration by temperature, humidity, light, shocks, physical damage, among others [5, 6]. Despite of several benefits that packaging industry offers to food, oxidation and microbial spoilage are the principal mechanisms that entail a great loss of fruits, vegetables,

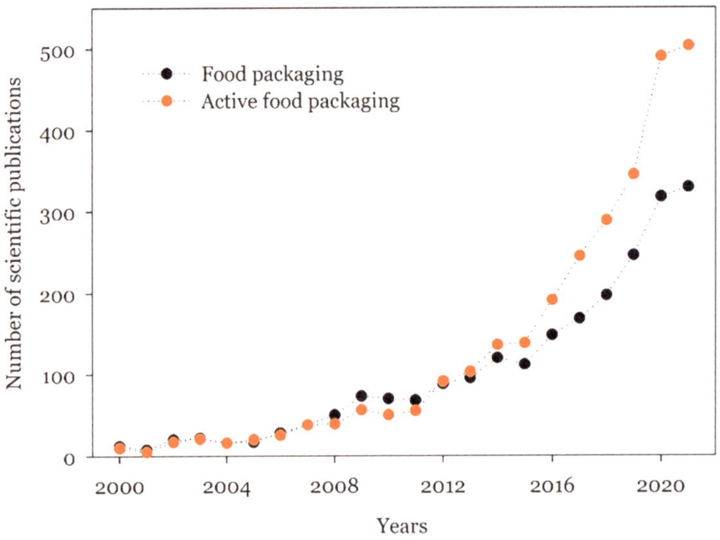

Figure 1.
Articles related to traditional food packaging and active food packaging in the last two decades (data obtained from Web of Science data basis).

meats, dairy and bakery products during their production, transport, processing, storage, and marketing [5]. In this context, during the last decades, the packaging area has centered its aims to the development of materials able to maintain or improve the properties of food, and therefore, extend its shelf-life. Several studies have developed different functional materials which in turn have allowed the surging of active food packaging in the last years. Thus, **Figure 1** shows a comparison between the number of indexed articles about "Food packaging" and "Active food packaging", published in the Web of Science (WOS) data basis. This figure shows the growing interest in food packaging research area during the last two decades, evidencing that active food packaging has been specifically leading during the last 5 years when compared to traditional food packaging published research.

2. Active food packaging technology

2.1 Definition

Active food packaging is considered a system of positive interaction between the food, the packaging material and the environment with the aim of preserving the properties of food and avoiding its deterioration during transport and storage [7]. In this way, compounds or substances with an active function (active compounds) have been incorporated into polymers in order to obtain active materials [6]. Natural and synthetic active compounds, such as plant extracts, essential oils, peptides, enzymes, organic acids, salts, metals ions, metal oxides, nanoparticles, among others, have afford to the packaging materials different functionalities, such as: (i) releasing/emitting of antioxidants, antimicrobial, sulfur dioxide, preservatives, ethanol and flavors; (ii) absorbing/scavenging of carbon dioxide, oxygen ethylene, flavors, moisture, UV light; and (iii) controlling the microbial, temperature and quality of foods [8, 9].

2.2 Formats and technologies

The design of active packaging systems has been mainly focused on the characteristics of the active compound and packaged food. **Figure 2** shows the five mechanisms for generating an active packaging system through [7, 10]:

Incorporation external devices (labels, pads or sachets loaded with some active compound) into the package able to release or absorb substances.

Coating of active compound onto inner surface of packaging material. This system is useful for heat-sensitive compounds or incompatible and immiscible with polymeric matrix.

Immobilization of active compound in the inner layer of packaging material through ion or covalent linkages. It is important the presence of functional groups on active compound and polymer to get the immobilization.

Direct addition of the active compound into the packaging material matrix [11].

By using polymers with some active function (e.g., chitosan) to develop composites or multilayer materials.

Most polymeric materials that have been part of above-mentioned active packaging systems have been obtained through traditional techniques, such as melting based processes (melt blending, hot pressing, cast extrusion, injection molding), casting, coating [12]. For example, the extrusion technique was recently used by producing active labels based on low-density polyethylene films loaded with different essential oils and vegetable oils [13]. In this work, fresh beef was packaged into commercial trays with modified atmosphere (70% O_2 + 30% CO_2), and the active labels were placed on the top of the tray in order to protect the food. Results

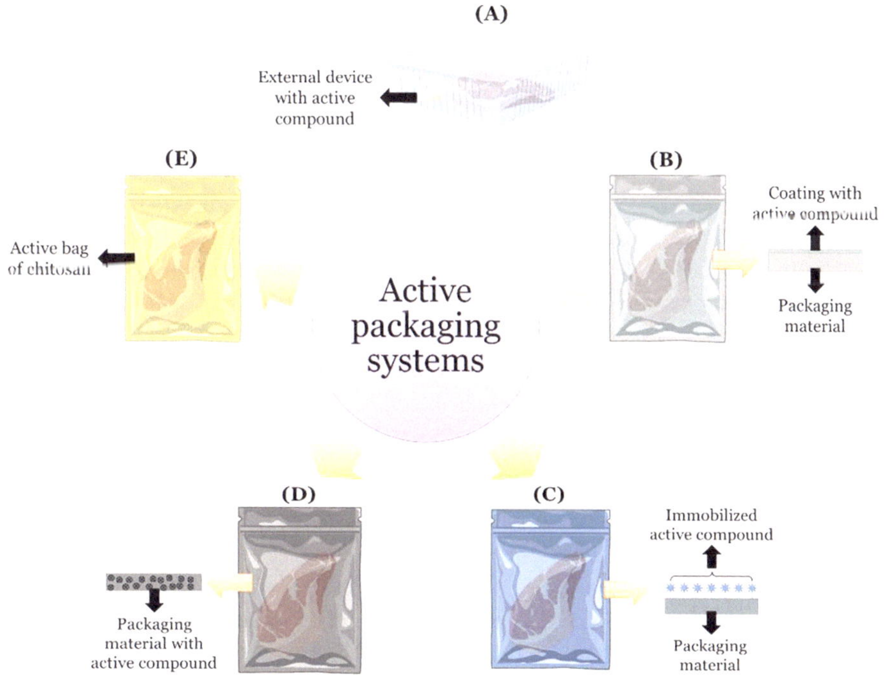

Figure 2.
Routes for obtaining an active packaging system.

demonstrated a great effectivity of active packaging system because the shelf-life of fresh meat was extended by 22%. On the other hand, casting technique has been also a useful tool for producing active food packaging systems. In this context, Sooch and Mann developed an active package from gelatin and copper nanoparticles doped with titanium dioxide. Tomatoes were wrapped with these packaging materials, and the shelf-life of vegetables was extended by 18 days [14]. Likewise, multilayer materials for active packaging of meat products have been also developed through coating technique. Bilayer films composed by poly(lactic) acid (PLA), as substrate, and chitosan or blends of chitosan/caseinate enriched with rosemary essential oil, as coating, were recently used for protecting fresh minced chicken meats. In this case, oxidation process and color changes of food were not shown for 14 days because the food was in direct contact with active materials [15].

On the other hand, non-traditional techniques as carbon dioxide supercritical impregnation and atomic layer deposition (ALD) have been also employed for developing active materials. However, the application of such technologies in food has been not already evaluated. Villegas et al. impregnated cinnamaldehyde through supercritical carbon dioxide into PLA film in order to obtain an antibacterial material. Active film showed a strong and effective antibacterial activity against *Staphylococcus aureus* and *Escherichia coli* [16]. Unlike supercritical impregnation, the combination of electrospinning and ALD process has allowed to produce metallic oxide nanostructures with antimicrobial properties that can be subsequently incorporated into polymers. In this context, nanotubes and spherical particles of titanium dioxide (TiO_2) and zinc oxide (ZnO) have been produced by this combination [17–19]. In all studies, metallic oxide nanostructures showed a high antimicrobial property against Gram-positive and Gram-negative bacteria.

As the technology has been progressed, in the last decade, a novel technique known as electrospinning has been used for developing polymeric materials that posteriorly play a fundamental role in the active packaging system.

3. Electrospinning

Electrospinning is an efficient and novel technique that consists in the application of an electric field to a polymeric solution in order to produce thinner structures known as “fibers” [11]. As **Figure 3** shows, the electrospinning is composed by three main components: (i) a high voltage source composed by two electrodes that are connected to the output of a metal needle and to collector, (ii) an injection pump that impulses the polymeric solution through plastic tube to metal needle, and (iii) a collector. To obtain the fibers, it is necessary that the surface tension of the drop formed in the tip of the needle be overcome by the force of the electric field. On this way, the polymeric solution is continuously stretched to produce a jet with a conical structure named “Taylor’s cone”, where the solvent is evaporated to obtain the fibers [12].

Characteristics of the fibers depend on the control of the electrospinning parameters. In this context, a change of the properties of the polymeric solution (polymer concentration, viscosity, electrical conductivity, type of solvent) or the operational parameters (flow rate, voltage and the height also known as the distance between the tip of the needle and the collector) can affect the size and morphology of the fibers [20]. This fact will be detailed in the following section.

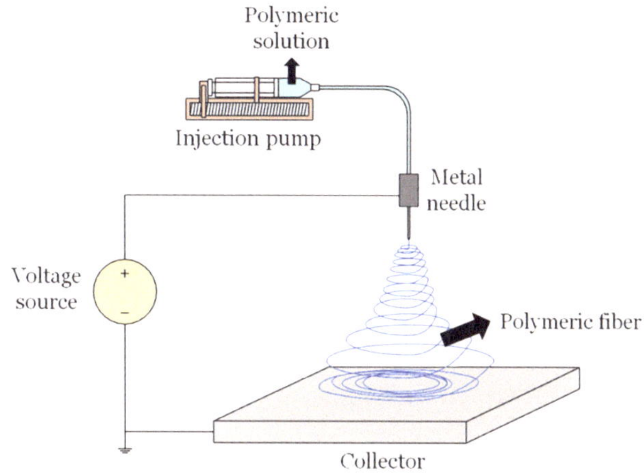

Figure 3.
Electrospinning system.

3.1 Influence of the properties of the polymeric solution

3.1.1 Polymer concentration

The concentration of the polymer is one of the main parameters that affects the size and the morphology of the fibers, and it is closely related to the viscosity of the solution. If the solution concentration is very low, the jet cannot be continuously stretched, and thus, uniform fibers cannot be obtained. This fact in turn can produce a decrease in the diameter of the fibers and the presence of beads in their surface. On the contrary, an increase of the concentration can result in thicker structures, and in some cases, the non-formation of the fibers due to high viscosity of the solution. A recent study evidenced the decrease of the diameter and the presence of beads in polycaprolactone (PCL) nanofibers when the polymeric concentration decreased from 13 to 8 wt% [21]. A similar result was also obtained in the processing of cellulose acetate (CA) and poly(vinyl chloride) (PVC) nanofibers. In both cases, the use of CA and PVC concentrations at 12 wt% resulted in thinner and beaded fibers, while an increase to 16 wt% produced structures with smooth surfaces and large diameters [22].

3.1.2 Electrical conductivity

The increase of electrical conductivity in the polymeric solution has been mainly related with the increase of the concentration of the polymer, favoring the electrospinning process and the formation of the fibers. For example, gelatin nanofibers were only obtained with acid solutions at high gelatin concentrations. The low electrical conductivity obtained with the lowest gelatin solution (7 wt%) did not allow to produce fibers. On the contrary, the increase of polymeric solution concentration to 20 wt% produced an increase of electrical conductivity and the formation of nanofibers [23].

3.1.3 Type of solvent

The type of solvent has an important role during the electrospinning process. It is stretched related with the surface tension of the solution, and several investigations have shown changes of this parameter by the use of different solvents or the mix of them. This fact in turn has derived in different morphologies and sizes of fibers. Thus, the combination of formic acid (FA) and dichloromethane (DCM) at different ratios as solvent system for producing PCL nanofibers produced changes on the diameter of the structures. A higher amount of FA produced an increase of electrical conductivity solution, while an increase of the amount of DCM increased its surface tension. These effects in turn produced fibers with diameters between 1.5 μm and 220 nm [21]. Likewise, the use of the following solvent systems: acetone/N,N-dimethylacetamide, acetone/N,N-dimethylformamide and tetrahydrofuran/dimethylformamide at different ratios changed the morphology of PCL nanofibers. Beaded, beaded-free, thin, thick and smooth nanofibers were obtained with these different combinations, and this fact could be associated to the changes on the surface tension of polymeric solutions [22].

3.2 Influence of operational parameters

3.2.1 Flow rate

The flow rate is considered a key parameter because controls the diameter of the fibers, the trajectory of jet, initial droplet shape, maintenance of Taylor's cone and the collection area. A high flow rate produces larger droplets in the tip of the needle and favors the formation of thicker and beaded fibers due to minimum solvent evaporation. On the contrary, a low flow rate facilitates the evaporation of solvent, obtaining uniform and smooth structures [24]. This effect has been observed when a poly(vinylidenefluoridene-hexafluoropropylene) solution was electrospun at different flow rates. An increase of flow rate from 0.1 to 0.7 mL/h produced an increase of fiber diameter [25]. The same result was also obtained when the flow rate of a poly(vinyl alcohol) (PVOH) solution increased from 0.75 to 1.5 mL/h [19].

3.2.2 Collection distance

The distance between the needle and the collector has a great effect on the diameter and the shape of the fibers. A small distance avoids the total evaporation of the solvent, and thus, the collection of wet and thicker fibers can occur. In this way, the presence of beads or the formation of ribbon-flat fibers have been the most common results. On the contrary, a high distance can improve the stretching of the jet and favor the formation of uniform and thinner structures. Furthermore, the increase of distance can also increase the needed voltage to produce the fibers [19, 26]. A recent study obtained thinner polyacrylonitrile nanofibers when increased the collection distance from 15 to 45 cm [27]. Similarly, the increase of collection distance from 8.5 to 10 cm produced smaller diameter in PVOH nanofibers [19].

3.2.3 Voltage

Despite of several studies have demonstrated that voltage has a minimum impact on the diameter and morphology of fibers, it is an important parameter to be considered during processing of electrospun materials. This fact is due to its increase can produce change the diameter of fibers and produce the presence or absence of beads in the structures [26]. Some studies have evidenced the effect of this parameter on such characteristics [23, 27].

3.3 Active electrospun materials

As it was earlier mentioned, electrospinning is able to produce smaller and thinner mats composed by fibers with high aspect ratio. In order to functionalize these mats, active compounds can be incorporated into them. These materials in turn could be converted to active packaging materials or be part of an active packaging system, exhibiting the following advantages [11]:

- Lower amount of required active compound due to small sizes of fibers,
- Good distribution of the active compound,
- High surface-to-volume ratio with tailored thicknesses due to the high versatility of this technique.

Despite of active electrospun materials are an excellent alternative to develop active packaging materials in comparison with other traditional techniques (melting processes, coating, casting), their processing remain one of the main challenges. This fact is mainly because the addition of the active compound can produce changes in the properties of the polymeric solution, and therefore, the electrospinning process can be affected. In order to deepen in the topic, the following section will discuss the effect of adding active compounds in the development electrospun materials.

3.4 Influence of active compounds during electrospinning process

The incorporation of an active compound to the polymeric solution to be electrospun can mainly change their physicochemical properties and affect the development of electrospun packaging material as follows.

3.4.1 Viscosity

The viscosity of the polymeric solution is related to the concentration and molecular weight of the polymer. A polymeric solution with an optimum viscosity can be electrospun in order to obtain the active mat. On the contrary, solutions with very high or low viscosities affect the electrospinning process and the obtaining of homogenous fibers [28]. Depending on the type of the active compound, the viscosity of the polymeric solution can be increased or decreased, and the morphology and diameter of the fibers are affected. As is shown in **Table 1**, the use of essential oils has mainly produced an increase of solution viscosity, which has derived in an increase of fiber diameter. This effect has been mainly associated to a less stretching of the jet [30]. For example, the increasing addition of angelica essential oil into gelatin solutions produced high viscosities which resulted in thicker fibers [30]. Altan et al. also evidenced this same effect when active PLA fibers loaded with carvacrol were developed [29]. Likewise, thicker fibers of glycyrrhiza polysaccharide and polyoxide ethylene (PEO) loaded with tea tree essential oil encapsulated into gliadin nanoparticles were obtained by Cai et al. Like above mentioned studies, the active compound increased the solution viscosity and affected the diameter of the fibers [31].

On the other hand, the effect of incorporating plant extracts in the viscosity of the polymeric solutions has produced different behaviors. A clear tendency of increase or decrease of viscosity by the incorporation of plant extracts has been not obtained.

Polymeric solution	Active compound	Amount of active compound incorporated (%wt.)	Polymeric solution properties			Reference
			V	ST	EC	
Zein	Carvacrol	5–20	↑	—	—	[29]
PLA			↑	—	—	
Gelatin	Angelica essential oil	3–9	↑	—	↑	[30]
Glycyrrhiza polysaccharide and PEO	Tea tree oil-gliadin nanoparticles	2–10	↑	—	↑	[31]
PVOH	LEO and REO	10	↑	↓	↑	[32]
PCL	Sage extract	5–20	↓	C	↑	[33]
Zein	Tomato peel extract	5–20	↑	↓	↓	[34]
Zein	Jaboticaba peel extracts	5–11	↓	—	↑	[35]
PVOH	Coptis chinensis extract	5–15	↑	↑	—	[36]
Zein	TiO_2	1–5	↑	—	↓	[37]
PVOH and gum karaya	Ag	0.2–2	↑	—	↑	[38]
Ethylcelullose/gelatin	ZnO	1–2	↓	—	—	[39]
Amaranth protein and pullulan	Nisin	10 and 20	↓	—	↓	[40]
Gelatin	Nisin	0.5–3	M	—	↑	[41]
Chitosan/PVOH	Lysozyme	10–30	↓	M	↑	[42]

V = viscosity, ST = surface tension, EC = electrical conductivity. Trends: ↓ = decreased, ↑ = increased, M = maintained, C = constant.

Table 1.
Influence of the active compound incorporation on properties of polymeric solution.

A work about PCL fibers loaded with sage extract evidenced that the incorporation and increase of extract concentration in the polymeric solution decreased its viscosity and favored the production of thinner fibers [33]. Likewise, an increase of concentration of jaboticaba peel extract in a zein solution decreased its viscosity and the diameter of the fibers [35]. Meanwhile, an increase of concentration of tomato peel extract in a polymeric solution of zein produced an increase of its viscosity and the development of thicker fibers [34]. Yang et al. also obtained a similar result when *Coptis chinensis* extract was incorporated in a zein solution. Similarly, the extract increased the viscosity of the solution, but favored the decrease of diameter of the fibers [36].

The use of nanoparticles has also influenced the diameter and the morphology of the fibers, and this fact has been mainly related to changes in the solution viscosity. For example, although an increase of TiO_2 nanoparticles concentration into zein solutions caused high viscosity values, thinner fibers were obtained [37]. A similar effect was also observed in the development of PVOH/gum karaya nanofibers loaded with silver (Ag) nanoparticles, whose resulting nanofibers showed roughness in their surface, and this fact was attributed to an increase of the viscosity of the polymeric solution. Meanwhile, Liu et al. evidenced a contrary effect because the viscosity of a blend of ethylcellulose and gelatin decreased with the increase of ZnO nanoparticles concentration, but the diameter of fibers increased [39].

The development of fibers loaded with peptides and enzymes has also been an excellent strategy for producing potential active electrospun materials. Most studies have evidenced a decrease of the polymeric solution viscosity after their incorporation, which in turn has affected the size of fibers. A decrease of diameter of amaranth protein/pullulan nanofibers was evidenced by Soto et al. when nisin was incorporated to the polymeric solution. The increase of nisin content decreased the viscosity of the solution due to the molecular entanglement among the components, and this in turn influenced on the diameter of the fibers [40]. Likewise, a work about active nanofibers of chitosan/PVOH loaded with lysozyme showed a result similar. In this case, the incorporation and increase of the lysozyme concentration caused a decrease of viscosity, and therefore, a decrease of the nanofiber's diameter [42].

3.4.2 Surface tension

This parameter is closely related with the nature of the solvent. Surface tension shows the strong cohesiveness of the molecules in the solution, which allows the formation of the drop on the tip of the capillary before to be formed the jet by action of electric field [43]. A decrease of surface tension results on bead-free fibers while its increase produces instability of the jet and avoids the formation of the nanostructures [44]. As can be seen in **Table 1**, the incorporation of active compounds has mainly produced a decrease of surface tension of polymeric solutions due to their behavior as surfactant. Despite of this, the morphological properties of the fibers have been not affected. This effect was recently evidenced when essential oils of *Laurus nobilis* (LEO) and *Rosmarinus officinalis* (REO) were added to an aqueous PVOH solution [32]. The surface tension value was significantly lower than the solvent, and it decreased even more when the essential oils were added [32]. Horuz et al. also obtained a similar result in their study about active fibers of zein loaded with tomato peel extract. In this case, the addition of the extract decreased the surface tension without affecting the morphology of the fibers [34]. Likewise, the incorporation of sage extract into an organic PCL solution did not produce changes on the morphology of the fibers although its surface tension was decreased [33].

3.4.3 Electrical conductivity

Another of the key parameters for the development of electrospun active fibers is the electrical conductivity. The value in this parameter influences on the fiber's morphology. Generally, an increase of electrical conductivity facilities the elongation of the droplet and jet formation, therefore, thinner and bead-free fibers are reached [45]. **Table 1** shows that the incorporation of an active compound into a polymeric solution has generally produced an increase of its electrical conductivity. Despite of this, the morphology and the diameter of the fibers have shown different trends. For example, although the addition and increase of angelica essential oil concentration into gelatin solutions produced higher electrical conductivity values, an increase of diameter of the fibers was mainly influenced by the increase of the viscosity [30]. A similar result was recently obtained in a study about active nanofibers of glycyrrhiza polysaccharide and PEO loaded with tea tree essential oil encapsulated into gliadin nanoparticles. As the above-mentioned study, the active compound significantly increased the electrical conductivity of the polymeric solution. However, the increase of fiber diameter was attributed to the higher solution viscosities [31]. Meanwhile, a contrary effect was obtained in the study of Avila et al. In this case, the incorporation and increase of jaboticaba peel extract concentration into zein solutions increased their electrical conductivity, which favored the formation of thinner fibers [35]. The same effect was also evidenced in the electrospinning process of chitosan nanofibers loaded with lysozyme [42].

On the other hand, few studies have evidenced a decrease of this parameter when the active compound has been added to the polymeric solution, resulting in the formation of thicker structures. The increase of the diameter of the fibers due to the decrease of electrical conductivity has been explained because static charges of the solution are oriented to jet surface during the electrospinning process. In this way, the capacity of polymer solutions to be electrospun is increased [34]. For example, the adding of TiO_2 or tomato peel extract in a zein polymeric solution has evidenced such effect [34, 37].

3.5 Application of active electrospun materials on food packaging

Although electrospinning has been a novel tool for developing active food packaging materials, their application to industrial scale has been not still largely exploited. This fact has been mainly associated to its low technological readiness level (TRL) in the food area due to low-yield in the production of packaging materials [11]. However, the numerous researches about the application of this technology in real matrixes reveal their promissing and potential application in the food packaging area in the future. In this sense, the most recent developments that involve the application of active electrospun materials on different foodstuffs, the type of formats used for packaging and the principal assay experimental are shown in **Table 2**.

3.5.1 Meat products

In the food industry, chicken, meat and pork are considered the most appreciated and chosen meat products by the consumers due to a 15% preference of world population [68]. These food have been usually packaged into trays and bags through different systems in order to ensure their quality and safety [69]. Despite of this, microbial contamination, oxidation processes and loss of the sensory properties are still some of main issues that produce their spoilage. In order to avoid such complications, several developments based on the direct application of active electrospun materials as wraps or films has allowed their protection, as **Table 2** shows. In this way, the quality, safety and shelf-life of these products have been guaranteed during long-time in a cold storage condition. A wrapper active film based on PVOH electrospun nanofibers loaded with LEO and REO was used to protect chicken breast fillets against oxidation and microbial contamination during cold storage. The active mat inhibited their lipid oxidation up to 68% and decreased their microbial growth [32]. Likewise, an active electrospun wrap composed by PLA nanofibers loaded with inclusion complexes of γ-cyclodextrin/α-tocopherol was able to reduce the lipidic oxidation of beef up to 50% during its storage at 4°C for 21 days [48]. Li et al. also prepared active gelatin/zein fibers with resveratrol in order to wrap small pieces

Food category	Food Type	Applied format	Principal experimental assays	Reference
Meat products	Chicken	Wrapper film	Antimicrobial, antioxidant and sensory properties	[32, 46, 47]
	Meat	Film and wrapper film		[48–50]
	Pork	Wrapper film		[51–53]
Sea products	Fish	Wrapper film	Antimicrobial, antioxidant and sensory properties	[54–56]
	Seafood	Cover film		[57]
Fruits and vegetables	Fruits	Bag, wrapper film and sachet	Weight loss, firmness, antioxidant, antimicrobial, sensory properties, ripening rate and ethylene production	[58–61]
	Vegetables	Film and wrapper film		[37, 62, 63]
Bakery products	Bread	Sachet	Antimicrobial	[29, 64]
Dairy products	Cheese	Wrapper film	Antimicrobial and sensory properties	[65–67]

Table 2.
Application of active electrospun materials on different food matrices.

of pork. Samples were stored at 4°C and the active mat was able to extend their shelf-life by 3 days [51].

3.5.2 Freshwater and sea products

Fish and seafood are the main products obtained from salt and fresh water. They constitute a great source of animal protein and essential micronutrients, such as minerals, vitamins and essential fatty acids [70]. However, their high fatty content is one of the main reasons of decay, principally evidenced through changes of their sensory properties due to oxidation processes. Furthermore, their deterioration is also related to the presence and growth of pathogens and microorganisms. Therefore, and as **Table 2** shows, the latest developments about electrospun active materials have been mainly focused on the protection of freshwater and sea products against such deterioration processes. In order to achieve this, these products have been wrapped with developed active electrospun mats containing different polymers and active compounds. A work about the development of sodium caseinate/gelatin nanofibers loaded with essential oil of *Mentha spicata* L. and magnesium oxide nanoparticles evaluated their effectivity as fresh trout fillets packaging during cold storage for 13 days. This active material was able to reduce the oxidation in the fillets up to 93% and the presence of microbial population up to 5 log CFU/g [54]. In the same way, *Mentha longifolia* L. essential oil was encapsulated into carboxymethyl cellulose/gelatin nanofibers in order to produce a nanofibrous film able to improve the shelf-life of peeled freshwater prawns during 14 days in refrigeration. A low microbial growth and oxidation of the samples during the storage was reflected with the high sensory scores in terms of odor, color, texture, taste, and total acceptance [57]. Likewise, active electrospun mats composed by PVOH fibers loaded with poly(hexamethylene biguanide) hydrochloride or nisin have also protected these food against microbial growth [55, 56].

3.5.3 Fruits and vegetables

Another important group of food used for evaluating the effectivity of packaging systems are fruit and vegetables. Different strategies, such as the use of active sachets or packages with improve physical properties, have been applied for protecting them and extending their shelf-life [71, 72]. Despite of this, their weight loss, textural changes and fungal contamination are still the principal issues causing of their spoilage. In this sense, the most recent developments about wrapper films, bags, films and active sachets obtained through electrospinning have shown their potential application for active food packaging (**Table 2**). These materials have been mainly applied on strawberries, grapes, tomatoes and mushrooms, and promising results have been obtained. For example, an active electrospun sachet composed by PLA and PVOH/poly(ethylene glycol) nanofibers loaded with thyme essential oil was able to maintain the freshness and prolong the softening rate of strawberries stored at 20°C for 5 days [73]. Likewise, active zein fibers with allyl isothiocyanate were applied as a sachet into a package containing strawberries. In this case, the active sachet reduced the weight loss up to 36% and maintained their firmness during 15 days at 4°C [58]. On the other hand, two recent developments based on PVOH nanofibers loaded with inclusion complexes of β-cyclodextrin with cinnamon essential oil (CEO), and zein and ethyl celullose fibers loaded with CEO demonstrated their excellent effectivity during the storage of mushrooms. In both cases, the weight loss of the vegetables decreased, and therefore, their shelf-life was prolonged [62, 63].

3.5.4 Dairy and bakery products

As **Table 2** shows, cheese and bread have been the main food models used for evaluating the antimicrobial effectivity of active electrospun packaging materials. In order to avoid the bacterial and fungal contamination in this food category, wrapper films and sachets obtained from such materials have been developed. Cheese has been mainly wrapped with the active mats to avoid the growth of pathogens microorganisms. For example, a total inhibition of microbial growth of *Salmonella typhimurium*, *Listeria monocytogenes* and *Leuconostoc mesenteroides* was obtained by Soto et al. when cheese cubes stored at 4°C for 7 days were covered with active amaranth protein isolate and pullulan nanofibers loaded with nisin [65]. A study based on PEO nanofibers containing nisin-loaded poly-γ-glutamic acid/chitosan nanoparticles also evidenced a similar behavior. In this case, the strong antibacterial activity of the active mat against *L. monocytogenes* was observed on cheese samples stored at 4 and 20°C for 7 and 15 days, respectively [66].

On the other hand, active electrospun sachets have been the preferred format packaging for protecting bread samples. Contrary to cheese, the sachets have been applied without direct contact, avoiding mainly the fungal contamination. This fact was evidenced in the studies developed by Altan et al. and Fonseca et al. Both researches evidenced that the active sachets obtained from starch, zein and PLA nanofibers loaded with carvacrol were able to inhibit the growth of molds on bread stored at 25°C for 7 days [29, 64].

4. Conclusions

Electrospinning has been one of the most recent and novel technologies used in the packaging research area. This technique has allowed the development of electrospun mats composed by fibers with high aspect ratio. During their development, the modification of electrospinning parameters in turn have generated changes on the morphological characteristics of resulting fibers. Moreover, electrospun mats have been also functionalized through the incorporation of active compounds into polymeric solutions. This fact has eventually modified the viscosity, surface tension and electrical conductivity of the polymeric solutions, and therefore, the morphology and sizes of electrospun structures.

On the other hand, although this technique presents a current low technological readiness level in the food packaging area, the interest and projection of this technology to be applied is growing in an exponential way. This fact has been mainly evidenced by the diverse developments of active electrospun materials able to protect different products, such as meat, chicken, fish, pork, fruits, vegetables, bread, cheese, among others. Therefore, electrospun mats could be proposed as the new generation of materials to be used in the active food packaging.

Acknowledgements

The authors acknowledge the financial support of Agencia Nacional de Investigación y Desarrollo de Chile (ANID) through the Doctoral Scholarships CONICYT-PFCHA/Doctorado Nacional/2019-21190316, CONICYT-PFCHA/Doctorado Nacional/2019-21190326 and Fondecyt Regular 1200766.

Conflict of interest

The authors declare no conflict of interest.

Author details

Cristian Patiño Vidal[1,2*], Cristina Muñoz-Shugulí[1,2], Marcelo Patiño Vidal[3], María José Galotto[1,2,4] and Carol López de Dicastillo[1,2,4]

1 University of Santiago of Chile (USACH), Packaging Innovation Center (LABEN), Chile

2 University of Santiago of Chile (USACH), Center for the Development of Nanoscience and Nanotechnology (CEDENNA), Chile

3 Escuela Superior Politécnica de Chimborazo (ESPOCH), Faculty of Mechanics, Industrial Engineering School, Ecuador

4 University of Santiago of Chile (USACH), Technological Faculty, Food Science and Technology Department (DECYTAL), Chile

*Address all correspondence to: cristian.patino@usach.cl

References

[1] Piergiovanni L, Limbo S. Introduction to food packaging materials. In: SpringerBriefs in Molecular Science. Cham: Springer; 2016. pp. 1-3

[2] Siracusa V. Packaging material in the food industry. In: Antimicrobial Food Packaging. San Diego: Elsevier Inc; 2016. pp. 95-106. ISBN: 9780128007235

[3] Cheng H, Chen L, McClements DJ, Yang T, Zhang Z, Ren F, et al. Starch-based biodegradable packaging materials: A review of their preparation, characterization and diverse applications in the food industry. Trends in Food Science and Technology. 2021;**114**:70-82. DOI: 10.1016/j.tifs.2021.05.017

[4] Mangaraj S, Yadav A, Bal LM, Dash SK, Mahanti NK. Application of biodegradable polymers in food packaging industry: A comprehensive review. Journal of Packaging Technology and Research. 2019;**3**:77-96. DOI: 10.1007/s41783-018-0049-y

[5] Han JW, Ruiz-Garcia L, Qian JP, Yang XT. Food packaging: A comprehensive review and future trends. Comprehensive Reviews in Food Science and Food Safety. 2018;**17**:860-877

[6] Ribeiro-Santos R, Andrade M, de Melo NR, Sanches-Silva A. Use of essential oils in active food packaging: Recent advances and future trends. Trends in Food Science and Technology. 2017;**61**: 132-140. DOI: 10.1016/j.tifs.2016.11.021

[7] Almasi H, Jahanbakhsh Oskouie M, Saleh A. A review on techniques utilized for design of controlled release food active packaging. Critical Reviews in Food Science and Nutrition. 2020;**61**:1-21

[8] Vilela C, Kurek M, Hayouka Z, Röcker B, Yildirim S, Antunes MDC, et al. A concise guide to active agents for active food packaging. Trends in Food Science and Technology. 2018;**80**:212-222

[9] Chen X, Chen M, Xu C, Yam KL. Critical review of controlled release packaging to improve food safety and quality. Critical Reviews in Food Science and Nutrition. 2019;**59**:2386-2399

[10] Vasile C, Baican M. Progresses in food packaging, food quality, and safetyzcontrolled-release antioxidant and/or antimicrobial packaging. Molecules. 2021;**26**:1263. DOI: 10.3390/molecules26051263

[11] Patiño Vidal C, López de Dicastillo C, Rodríguez-Mercado F, Guarda A, Galotto MJ, Muñoz-Shugulí C. Electrospinning and cyclodextrin inclusion complexes: An emerging technological combination for developing novel active food packaging materials. Critical Reviews in Food Science and Nutrition. 2021

[12] Velásquez E, Patiño Vidal C, Rojas A, Guarda A, Galotto MJ, López de Dicastillo C. Natural antimicrobials and antioxidants added to polylactic acid packaging films. Part I: Polymer processing techniques. Comprehensive Reviews in Food Science and Food Safety. 2021;**20**:3388-3403. DOI: 10.1111/1541-4337.12777

[13] Wrona M, Silva F, Salafranca J, Nerín C, Alfonso MJ, Caballero MÁ. Design of new natural antioxidant active packaging: Screening flowsheet from pure essential oils and vegetable oils to ex vivo testing in meat samples. Food Control. 2021;**120**:107536. DOI: 10.1016/J.FOODCONT.2020.107536

[14] Sooch BS, Mann MK. Nanoreinforced biodegradable gelatin based active food

packaging film for the enhancement of shelf life of tomatoes (*Solanum lycopersicum* L.). Food Control. 2021;**130**: 108322. DOI: 10.1016/J.FOODCONT.2021.108322

[15] Fiore A, Park S, Volpe S, Torrieri E, Masi P. Active packaging based on PLA and chitosan-caseinate enriched rosemary essential oil coating for fresh minced chicken breast application. Food Packaging and Shelf Life. 2021;**29**:100708. DOI: 10.1016/J.FPSL.2021.100708

[16] Villegas C, Torres A, Rios M, Rojas A, Romero J, López de Dicastillo C, et al. Supercritical impregnation of cinnamaldehyde into polylactic acid as a route to develop antibacterial food packaging materials. Food Research International. 2017;**99**:650-659. DOI: 10.1016/j.foodres.2017.06.031

[17] López de Dicastillo C, Patiño C, Galotto MJ, Vásquez-Martínez Y, Torrent C, Alburquenque D, et al. Novel hollow titanium dioxide nanospheres with antimicrobial activity against resistant bacteria. Beilstein Journal of Nanotechnology. 2019;**10**:1716-1725. DOI: 10.3762/bjnano.10.167

[18] López de Dicastillo C, Patiño Vidal C, Falcó I, Sánchez G, Márquez P, Escrig J. Antimicrobial bilayer nanocomposites based on the incorporation of As-synthetized hollow zinc oxide nanotubes. Nanomaterials. 2020;**10**:503. DOI: 10.3390/nano10030503

[19] López de Dicastillo C, Patiño C, Galotto MJ, Palma JL, Alburquenque D, Escrig J. Novel antimicrobial titanium dioxide nanotubes obtained through a combination of atomic layer deposition and electrospinning technologies. Nanomaterials. 2018;**8**:128. DOI: 10.3390/nano8020128

[20] Muñoz-Shugulí C, Vidal CP, Cantero-López P, Lopez-Polo J. Encapsulation of plant extract compounds using cyclodextrin inclusion complexes, liposomes, electrospinning and their combinations for food purposes. Trends in Food Science and Technology. 2021;**108**:177-186

[21] Shahrousvand M, Haddadi-Asl V, Shahrousvand M. Step-by-step design of poly(ε-caprolactone) /chitosan/melilotus officinalis extract electrospun nanofibers for wound dressing applications. International Journal of Biological Macromolecules. 2021;**180**:36-50. DOI: 10.1016/J.IJBIOMAC.2021.03.046

[22] Tarus B, Fadel N, Al-Oufy A, El-Messiry M. Effect of polymer concentration on the morphology and mechanical characteristics of electrospun cellulose acetate and poly(vinyl chloride) nanofiber mats. Alexandria Engineering Journal. 2016;**55**:2975-2984. DOI: 10.1016/J.AEJ.2016.04.025

[23] Okutan N, Terzi P, Altay F. Affecting parameters on electrospinning process and characterization of electrospun gelatin nanofibers. Food Hydrocolloids. 2014;**39**:19-26. DOI: 10.1016/J.FOODHYD.2013.12.022

[24] Zargham S, Bazgir S, Tavakoli A, Rashidi AS, Damerchely R. The effect of flow rate on morphology and deposition area of electrospun Nylon 6 nanofiber. Journal of Engineered Fibers and Fabrics. 2012;**7**:42-49. DOI: 10.1177/155892501200700414

[25] Nawae S, Tohluebaji N, Putson C, Muensit N, Yuennan J. Effect of flow rate on the fabrication of P(VDF-HFP) nanofibers. Journal of Physics Conference Series. 2020;**1719**:012070. DOI: 10.1088/1742-6596/1719/1/012070

[26] Bhardwaj N, Kundu SC. Electrospinning: A fascinating fiber fabrication technique. Biotechnology

Advances. 2010;**28**:325-347. DOI: 10.1016/j.biotechadv.2010.01.004

[27] He H, Wang Y, Farkas B, Nagy ZK, Molnar K. Analysis and prediction of the diameter and orientation of AC electrospun nanofibers by response surface methodology. Materials and Design. 2020;**194**:108902. DOI: 10.1016/J.MATDES.2020.108902

[28] Ghorani B, Tucker N. Fundamentals of electrospinning as a novel delivery vehicle for bioactive compounds in food nanotechnology. Food Hydrocolloids. 2015;**51**:227-240. DOI: 10.1016/j.foodhyd.2015.05.024

[29] Altan A, Aytac Z, Uyar T. Carvacrol loaded electrospun fibrous films from zein and poly(lactic acid) for active food packaging. Food Hydrocolloids. 2018;**81**:48-59. DOI: 10.1016/j.foodhyd.2018.02.028

[30] Zhou Y, Miao X, Lan X, Luo J, Luo T, Zhong Z, et al. Angelica essential oil loaded electrospun gelatin nanofibers for active food packaging application. Polymers (Basel). 2020;**12**:299. DOI: 10.3390/polym12020299

[31] Cai M, Zhang G, Wang J, Li C, Cui H, Lin L. Application of glycyrrhiza polysaccharide nanofibers loaded with tea tree essential oil/gliadin nanoparticles in meat preservation. Food Bioscience. 2021;**43**:101270. DOI: 10.1016/J.FBIO.2021.101270

[32] Göksen G, Fabra MJ, Pérez-Cataluña A, Ekiz HI, Sanchez G, López-Rubio A. Biodegradable active food packaging structures based on hybrid cross-linked electrospun polyvinyl alcohol fibers containing essential oils and their application in the preservation of chicken breast fillets. Food Packaging and Shelf Life. 2021;**27**:100613. DOI: 10.1016/J.FPSL.2020.100613

[33] Salević A, Prieto C, Cabedo L, Nedović V, Lagaron J. Physicochemical, antioxidant and antimicrobial properties of electrospun poly(ε-caprolactone) films containing a solid dispersion of sage (*Salvia officinalis* L.) extract. Nanomaterials. 2019;**9**:270. DOI: 10.3390/nano9020270

[34] Horuz Tİ, Belibağlı KB. Nanoencapsulation of carotenoids extracted from tomato peels into zein fibers by electrospinning. Journal of the Science of Food and Agriculture. 2019;**99**:759-766. DOI: 10.1002/JSFA.9244

[35] Avila LB, Fontes MRV, Zavareze ER, Moraes CC, Morais MM, da Rosa GS. Recovery of bioactive compounds from jaboticaba peels and application into zein ultrafine fibers produced by electrospinning. Polymer. 2020;**12**:2916. DOI: 10.3390/POLYM12122916.

[36] Yang SB, Kim EH, Kim SH, Kim YH, Oh W, Lee J-T, et al. Electrospinning fabrication of poly(vinyl alcohol)/coptis chinensis extract nanofibers for antimicrobial exploits. Nanomaterials. 2018;**8**:734. DOI: 10.3390/NANO8090734

[37] Böhmer-Maas BW, Fonseca LM, Otero DM, da Rosa Zavareze E, Zambiazi RC. Photocatalytic zein-TiO2 nanofibers as ethylene absorbers for storage of cherry tomatoes. Food Packaging and Shelf Life. 2020;**24**:100508. DOI: 10.1016/J.FPSL.2020.100508

[38] Padil VVT, Nguyen NHA, Ševcu A, Černík M. Fabrication, characterization, and antibacterial properties of electrospun membrane composed of gum karaya, polyvinyl alcohol, and silver nanoparticles. Journal of Nanomaterials. 2015;**2015**:1-10. DOI: 10.1155/2015/750726

[39] Liu Y, Li Y, Deng L, Zou L, Feng F, Zhang H. Hydrophobic ethylcellulose/

gelatin nanofibers containing zinc oxide nanoparticles for antimicrobial packaging. Journal of Agricultural and Food Chemistry. 2018;**66**:9498-9506. DOI: 10.1021/ACS.JAFC.8B03267

[40] Soto KM, Hernández-Iturriaga M, Loarca-Piña G, Luna-Bárcenas G, Gómez-Aldapa CA, Mendoza S. Stable nisin food-grade electrospun fibers. Journal of Food Science and Technology. 2016;**53**(10):3787-3794. DOI: 10.1007/S13197-016-2365-Y

[41] Dheraprasart C, Rengpipat S, Supaphol P, Tattiyakul J. Morphology, release characteristics, and antimicrobial effect of nisin-loaded electrospun gelatin fiber mat. Journal of Food Protection. 2009;**72**:2293-2300. DOI: 10.4315/0362-028X-72.11.2293

[42] Charernsriwilaiwat N, Opanasopit P, Rojanarata T, Ngawhirunpat T. Lysozyme-loaded, electrospun chitosan-based nanofiber mats for wound healing. International Journal of Pharmaceutics. 2012;**427**:379-384. DOI: 10.1016/J.IJPHARM.2012.02.010

[43] Amariei N, Manea LR, Bertea AP, Bertea A, Popa A. The influence of polymer solution on the properties of electrospun 3D nanostructures. IOP Conference Series: Materials Science and Engineering. 2017;**209**:012092. DOI: 10.1088/1757-899X/209/1/012092

[44] Ibrahim HM, Klingner A. A review on electrospun polymeric nanofibers: Production parameters and potential applications. Polymer Testing. 2020;**90**:106647.DOI: 10.1016/J.POLYMERTESTING.2020.106647

[45] Mercante LA, Scagion VP, Migliorini FL, Mattoso LHC, Correa DS. Electrospinning-based (bio)sensors for food and agricultural applications: A review. TrAC Trends in Analytical Chemistry. 2017;**91**:91-103. DOI: 10.1016/J.TRAC.2017.04.004

[46] Surendhiran D, Cui H, Lin L. Encapsulation of phlorotannin in alginate/PEO blended nanofibers to preserve chicken meat from Salmonella contaminations. Food Packaging and Shelf Life. 2019;**21**:100346. DOI: 10.1016/j.fpsl.2019.100346

[47] Lin L, Zhu Y, Cui H. Electrospun thyme essential oil/gelatin nanofibers for active packaging against *Campylobacter jejuni* in chicken. LWT. 2018;**97**:711-718. DOI: 10.1016/j.lwt.2018.08.015

[48] Aytac Z, Keskin NOS, Tekinay T, Uyar T. Antioxidant α-tocopherol/γ- cyclodextrin-inclusion complex encapsulated poly(lactic acid) electrospun nanofibrous web for food packaging. Journal of Applied Polymer Science. 2017;**134**:44858. DOI: 10.1002/app.44858

[49] Cui H, Bai M, Lin L. Plasma-treated poly(ethylene oxide) nanofibers containing tea tree oil/beta-cyclodextrin inclusion complex for antibacterial packaging. Carbohydrate Polymers. 2018;**179**:360-369. DOI: 10.1016/j.carbpol.2017.10.011

[50] Aytac Z, Ipek S, Durgun E, Tekinay T, Uyar T. Antibacterial electrospun zein nanofibrous web encapsulating thymol/cyclodextrin-inclusion complex for food packaging. Food Chemistry. 2017;**233**:117-124. DOI: 10.1016/j.foodchem.2017.04.095

[51] Li L, Wang H, Chen M, Jiang S, Cheng J, Li X, et al. Gelatin/zein fiber mats encapsulated with resveratrol: Kinetics, antibacterial activity and application for pork preservation. Food Hydrocolloids. 2020;**101**:105577. DOI: 10.1016/J.FOODHYD.2019.105577

[52] Yang Y, Shi Y, Cao X, Liu Q, Wang H, Kong B. Preparation and functional

properties of poly(vinyl alcohol)/ethyl cellulose/tea polyphenol electrospun nanofibrous films for active packaging material. Food Control. 2021;**130**:108331. DOI: 10.1016/J.FOODCONT.2021.108331

[53] Wen P, Zhu DH, Feng K, Liu FJ, Lou WY, Li N, et al. Fabrication of electrospun polylactic acid nanofilm incorporating cinnamon essential oil/β-cyclodextrin inclusion complex for antimicrobial packaging. Food Chemistry. 2016;**196**:996-1004. DOI: 10.1016/j.foodchem.2015.10.043

[54] Eghbalian M, Shavisi N, Shahbazi Y, Dabirian F. Active packaging based on sodium caseinate-gelatin nanofiber mats encapsulated with *Mentha spicata* L. essential oil and MgO nanoparticles: Preparation, properties, and food application. Food Packaging and Shelf Life. 2021;**29**:100737. DOI: 10.1016/J.FPSL.2021.100737

[55] Zhang Y, Yang L, Dong Q, Li L. Fabrication of antibacterial fibrous films by electrospinning and their application for Japanese sea bass (*Lateolabrax japonicus*) preservation. LWT. 2021;**149**:111870. DOI: 10.1016/J.LWT.2021.111870

[56] Ozogul Y, Yuvka İ, Ucar Y, Durmus M, Kösker AR, Öz M, et al. Evaluation of effects of nanoemulsion based on herb essential oils (rosemary, laurel, thyme and sage) on sensory, chemical and microbiological quality of rainbow trout (*Oncorhynchus mykiss*) fillets during ice storage. LWT-Food Science and Technology. 2017;**75**:677-684. DOI: 10.1016/J.LWT.2016.10.009

[57] Shahbazi Y, Shavisi N, Karami N, Lorestani R, Dabirian F. Electrospun carboxymethyl cellulose-gelatin nanofibrous films encapsulated with *Mentha longifolia* L. essential oil for active packaging of peeled giant freshwater prawn. LWT. 2021;**152**:112322. DOI: 10.1016/J.LWT.2021.112322

[58] Colussi R, Ferreira da Silva WM, Biduski B, Mello El Halal SL, da Rosa Zavareze E, Guerra Dias AR. Postharvest quality and antioxidant activity extension of strawberry fruit using allyl isothiocyanate encapsulated by electrospun zein ultrafine fibers. LWT. 2021;**143**:111087. DOI: 10.1016/J.LWT.2021.111087

[59] Min T, Sun X, Zhou L, Du H, Zhu Z, Wen Y. Electrospun pullulan/PVA nanofibers integrated with thymol-loaded porphyrin metal–organic framework for antibacterial food packaging. Carbohydrate Polymers. 2021;**270**:118391. DOI: 10.1016/J.CARBPOL.2021.118391

[60] Li T, Liu Y, Qin Q, Zhao L, Wang Y, Wu X, et al. Development of electrospun films enriched with ethyl lauroyl arginate as novel antimicrobial food packaging materials for fresh strawberry preservation. Food Control. 2021;**130**:108371. DOI: 10.1016/J.FOODCONT.2021.108371

[61] Zhu Z, Zhang Y, Shang Y, Wen Y. Electrospun nanofibers containing TiO2 for the photocatalytic degradation of ethylene and delaying postharvest ripening of bananas. Food and Bioprocess Technology. 2018;**122**(12):281-287. DOI: 10.1007/S11947-018-2207-1

[62] Pan J, Ai F, Shao P, Chen H, Gao H. Development of polyvinyl alcohol/β-cyclodextrin antimicrobial nanofibers for fresh mushroom packaging. Food Chemistry. 2019;**300**:125249. DOI: 10.1016/j.foodchem.2019.125249

[63] Niu B, Zhan L, Shao P, Xiang N, Sun P, Chen H, et al. Electrospinning of zein-ethyl cellulose hybrid nanofibers

with improved water resistance for food preservation. International Journal of Biological Macromolecules. 2020;**142**:592-599. DOI: 10.1016/j.ijbiomac.2019.09.134

[64] Fonseca LM, Souza EJD, Radünz M, Gandra EA, Zavareze ER, Dias ARG. Suitability of starch/carvacrol nanofibers as biopreservatives for minimizing the fungal spoilage of bread. Carbohydrate Polymers. 2021;**252**:117166. DOI: 10.1016/J.CARBPOL.2020.117166

[65] Soto KM, Hernández-Iturriaga M, Loarca-Piña G, Luna-Bárcenas G, Mendoza S. Antimicrobial effect of nisin electrospun amaranth: Pullulan nanofibers in apple juice and fresh cheese. International Journal of Food Microbiology. 2019;**295**:25-32. DOI: 10.1016/J.IJFOODMICRO.2019.02.001

[66] Cui H, Wu J, Li C, Lin L. Improving anti-listeria activity of cheese packaging via nanofiber containing nisin-loaded nanoparticles. LWT-Food Science and Technology. 2017;**81**:233-242. DOI: 10.1016/J.LWT.2017.04.003

[67] Tayebi-Moghaddam S, Khatibi R, Taklavi S, Hosseini-Isfahani M, Rezaeinia H. Sustained-release modeling of clove essential oil in brine to improve the shelf life of Iranian white cheese by bioactive electrospun zein. International Journal of Food Microbiology. 2021;**355**: 109337. DOI: 10.1016/J.IJFOODMICRO. 2021.109337

[68] Organisation for Economic Co-operation Development-Food and Agriculture Organization. OECD-FAO Agricultural Outlook 2021-2030; 2021

[69] Papadochristopoulos A, Kerry JP, Fegan N, Burgess CM, Duffy G. Natural anti-microbials for enhanced microbial safety and shelf-life of processed packaged meat. Food. 2021;**10**:1598. DOI: 10.3390/FOODS10071598

[70] Sampantamit T, Ho L, Lachat C, Hanley-Cook G, Goethals P. The contribution of Thai fisheries to sustainable seafood consumption: National trends and future projections. Food. 2021;**10**:880. DOI: 10.3390/FOODS10040880

[71] Deshwal GK, Tiwari S, Panjagari NR, Masud S. Active packaging of fruits and vegetables: Quality preservation and shelf-life enhancement. In: Packaging and Storage of Fruits and Vegetables. New York: Apple Academic Press; 2021. pp. 109-131. ISBN 9781003161165

[72] Basumatary IB, Mukherjee A, Katiyar V, Kumar S. Biopolymer-based nanocomposite films and coatings: Recent advances in shelf-life improvement of fruits and vegetables. Critical Reviews in Food Science and Nutrition. 2020. DOI: 10.1080/10408398.2020.1848789

[73] Min T, Sun X, Yuan Z, Zhou L, Jiao X, Zha J, et al. Novel antimicrobial packaging film based on porous poly(lactic acid) nanofiber and polymeric coating for humidity-controlled release of thyme essential oil. LWT. 2021;**135**:110034. DOI: 10.1016/j.lwt.2020.110034

Chapter 2

Electrospun Polymeric Substrates for Tissue Engineering: Viewpoints on Fabrication, Application, and Challenges

Azadeh Izadyari Aghmiuni, Arezoo Ghadi, Elmira Azmoun, Niloufar Kalantari, Iman Mohammadi and Hossein Hemati Kordmahaleh

Abstract

Electrospinning is the technique for producing nonwoven fibrous structures, to mimic the fabrication and function of the native extracellular matrix (ECM) in tissue. Prepared fibrous with this method can act as potential polymeric substrates for proliferation and differentiation of stem cells (with the cellular growth pattern similar to damaged tissue cells) and facilitation of artificial tissue remodeling. Moreover, such substrates can improve biological functions, and lead to a decrease in organ transplantation. In this chapter, we focus on the fundamental parameters and principles of the electrospinning technique to generate natural ECM-like substrates, in terms of structural and functional complexity. In the following, the application of these substrates in regenerating various tissues and the role of polymers (synthetic/natural) in the formation of such substrates is evaluated. Finally, challenges of this technique (such as cellular infiltration and inadequate mechanical strength) and solutions to overcome these limitations are studied.

Keywords: electrospinning process, parameters of electrospinning technique, synthetic/natural polymers, electrospun scaffolds, Tissue engineering

1. Introduction

Nowadays, tissue engineering is known as a multi-disciplinary science that leads to the regeneration of damaged/lost tissues via combining the cell and scaffold [1, 2]. In this technique, the engineered scaffolds act as micro/nano/smart environments for improving the interactions of cell-scaffold and cellular functions (such as proliferation, adhesion, differentiation, and growth) [3]. The methods that can be led to the design of scaffolds similar to extracellular matrices (ECM), with the properties of mechanical/biochemical support from the cells and imitation of architecture/structure and function of the tissues, play an important role in this field [4–7].

Electrospinning is one of the unique approaches in this field that lead to the production of polymeric fibers with interconnecting pores and the opportunity to control the network and morphology of scaffolds, especially in bio-polymers processing. Such that, the polymeric solutions with low viscosity lead to shorter and finer scaffolds/substrates, while, more viscous solutions provide a relatively continuous scaffold. However, the morphology and diameter of the produced fibers depend on the processing conditions and the type of the polymer [8].

These conditions can be provided via controlling parameters of the solution, process, and ambient, such as an electric-field application, distance between the needle and collector, needle diameter, and flow rate, solution conductivity, the concentration of the polymeric solution, materials molecular weight, and solution viscosity [9–13].

In recent years, many studies have been carried out on electrospun substrates and their functions for tissue engineering applications, such as the regeneration of the blood vessels, skin tissue, cartilage, and bone, as well as muscle [14–23]. However, a better understanding of cellular responses to sophisticated structures derived from this technique can be effective in reaching ECM-liked substrates.

Hence, in this chapter, we discuss the principles of the electrospinning technique to analyze these sophisticated structures and generate natural ECM-like substrates. In the following, the application of these substrates in regenerating various tissues and

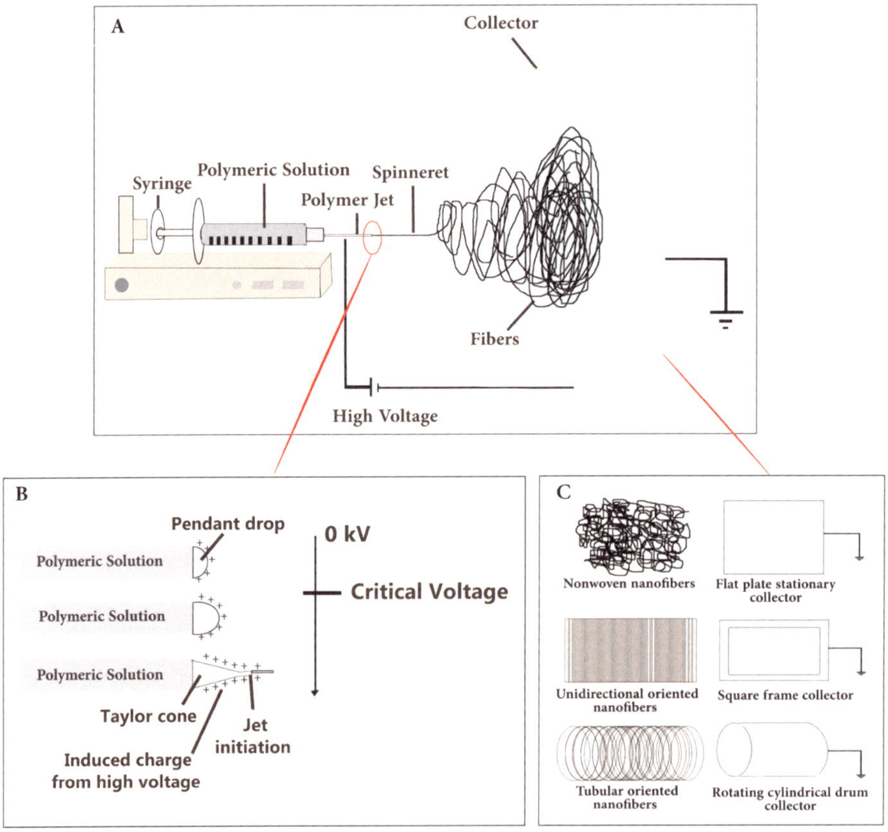

Figure 1.
(A) Schematic of electrospinning process, (B) Taylor cone formation via increasing the voltage, (C) the fibers derived from electrospinning process with different collectors.

the role of polymers (synthetic/natural) in the formation of such substrates is evaluated. Finally, challenges of this technique and solutions to overcome these limitations are studied.

2. Electrospinning process

In this simple technique (spinning technique), a very high electrical field (high voltage), in the range of 10–50 kV, is applied for accelerating the charged polymer jet and producing ultrafine fibers (**Figure 1A**). Moreover, the electrospinning machine includes a syringe pump along with a syringe (with a metallic needle) that is loaded with the polymeric solutions, so that the tip needle is attached to one of the negative or positive terminals of the high-voltage electrical field, and pendant-shaped droplets of the polymer solution are held through surface tension. Notably, the needle tip is usually attached to the positive terminal of the electrical field [24]. In these conditions, the increase of the voltage is led to the formation of the Taylor cone in the needle tip [25]. Afterward, increased voltage leads to the creation of the critical value above which the electrostatic forces can overcome the surface tension forces so that it results in ejecting out the fine-jet of the solutions from the tip of the Taylor cone (**Figure 1B**). In the following, solvents are evaporated at a low boiling point due to contact with the atmosphere and subsequently the charged polymeric strands deposited on the collector. The collectors can play a crucial impact in reaching the various structures of the scaffold. Such that, unidirectionally oriented nanofibers, aligned nanofibers, and non-woven nanofibers can be designed by square frame collector, rotating collector-drum, and flat plate stationary collector [26], respectively (**Figure 1C**).

3. Electrospinning parameters

The parameters of the electrospinning process play an important role in understanding the nature of this process and conversing polymeric solutions into nanofibers. Indeed, control of these parameters can be led to electrospun fibers with a desired morphology and diameter. Hence, this section has been focused on these parameters and their influence on the properties of produced fibers. The mentioned parameters have been listed in **Table 1**.

4. Polymers for electrospun fibers

Nowadays, natural and synthetic polymers are widely used in the design of electrospun scaffolds for tissue engineering applications [6, 51–56]. In this field, synthetic polymers possess high flexibility in the electrospinning process and can provide fibers with better mechanical properties [56, 57]. Although, these polymers are also highly cost-effective than bio/natural polymers, however, a comparison of these two polymers indicates that synthetic polymers lack bioactivity and need more modification to improve biological properties [53, 56, 58]. In contrast, natural polymers possess the properties of inherent bioactive and can be led to an increase in the interactions of scaffold-cell and cell–cell (i.e., adhesion, proliferation, differentiation) [3]. These polymers have a relatively low immune response (in terms of chemical degradation) and can provide a structure similar to native ECM. In recent years, more

Parameters	Details	Description	References
Solution	Concentration	There are four critical concentrations for this parameter: a. Very low concentrations that led to polymeric micro/nanoparticles b. Little higher concentrations that led to the mixture of bead and fiber. c. Suitable concentrations that led to smooth nanofibers. d. Very high concentrations that led to helix-shaped micro-ribbons. Notably, adjusting the concentration can tune the polymeric solution viscosity.	[27–30]
	Molecular weight (M_w)	This parameter can reflect the entanglement of polymer chains in the polymeric solution. However, when sufficient intermolecular interactions are supplied via oligomers, the control of Mw is not essential for electrospinning processing. Generally, in constant concentrations, the reduction of molecular weight is led to the formation of beads, rather than smooth fibers. Likewise, the increase in the polymer molecular weight provides smooth fibers. Notably, the too-high molecular weights can form micro-ribbon and some patterned fibers, even in low concentrations.	[31–33]
	Viscosity	The selection of suitable viscosity plays a critical impact in the determination of the fiber morphologies, so that, the very low viscosity cannot provide the continuous/smooth fibers. Moreover, very high viscosity leads to the harder ejection of the jet from polymeric solutions. The concentration and molecular weight are the effective parameters on the viscosity, such that, adjustment of these two parameters can lead to adjustment solution viscosity.	[34–37]
	Surface tension	This parameter is considered as the solvent composition function in the polymeric solutions and plays a crucial impact in the electrospinning process. Indeed, surface tension can act as a dominant factor for the low viscosity of the solutions and lead to the bead/beaded fiber formation. In these conditions, the controlled constant concentration can reduce the surface tension and beaded fibers convert into smooth fibers. Moreover, given that different solvents possess different surface tensions, the change of the mass ratio of the solvent mix can lead to an adjustment in solution viscosity and surface tension.	[30, 38–40]
	Conductivity/ surface charge density	This parameter is usually determined and tuned via the type of solvent (such as formic acid), polymer (such as PCL), and salts (such as KH_2PO_4, NaCl). Accordingly, the use of ionic salts and organic acids leads to the production of nanofibers with small diameters and solutions with high conductivity, respectively. Generally, biopolymers are led to more poor fibers compared to synthetic polymers; due to their polyelectrolytic nature that led to an increase in the carried charge by ions, and higher tension by the electric field	[41, 42]
Processing	Voltage	In this parameter, the voltages higher than the threshold voltage can lead to the ejection of the charged jet from Taylor Cone. Moreover, the applied voltage plays an important role in adjusting the diameter of fibers, so that, based on the studies, high voltage can facilitate the	[27, 40, 43–45]

Parameters	Details	Description	References
		formation of fibers with a large diameter. There are also several reports that show the increased voltages and consequently increase of electrostatic-repulsive forces of the charged jets can provide the fibers with a smaller diameter. Moreover, some studies also indicated that higher voltages can increase the formation of beads	
	Flow rate of polymeric solutions within the syringe	Based on the reports, a lower flow rate can be more effective in the electrospinning process due to the better polarization of polymeric solution. Indeed, in very high flow rates, bead fibers (thick diameters) form rather than the smooth fibers (thin diameters) that can be due to lower stretching force and the short time of drying polymeric solution (prior to contacting the collector)	[44, 45]
	Collector	The collectors (mainly Al foil) are usually used as conductive substrates for collecting the charged fiber. In this field, there are many collectors such as wire meshes, parallel or gridded bars, pins, rotating rods or wheels, liquid baths, grids, etc.	[46–49]
	Distance between the collector and the syringe tip	This parameter also plays a crucial impact in controlling the diameter of fibers and their morphology [12]. Such that, in the too-short distances, the fibers do not possess enough time for solidifying prior to contacting the collector. Likewise, too-long distances lead to the formation of bead fibers. Indeed, this parameter is known as the physical factor of electrospun fibers and plays an important role in drying solvent	[43, 45]
Ambient	Temperature and humidity	The temperature and humidity are known as ambient parameters that can be effective on the diameter of fibers and their morphology. Based on the reports, the increase in process temperature can lead to the formation of fibers with a thinner diameter, as well as low humidity can be resulted in increasing solvent evaporation velocity and drying the solvent. Likewise, higher humidity can form the fibers with thick diameters due to the neutralization of charge on the jet and smaller stretching force	[45, 50]

Table 1.
The important parameters in the electrospinning process, as well as the morphology of electrospun fibers.

than 200 natural/synthetic/copolymer/hybrid polymers have been designed and studied to obtain electrospun scaffolds with suitable physicomechanical and biological properties for use in tissue engineering [16, 17, 59–65]. Some of these polymers and their applications have been listed in **Table 2**.

4.1 Natural polymers

4.1.1 Collagen

Collagen is one of the main components of the native ECM with diameters in the range of 10–500 nm that plays an important role in providing mechanical strength of tissue and stimulating cell attachment and its proliferation [67, 83, 84]. Generally, type I collagen is the most common type of this protein in the dermis (70–80%),

The type of polymer	Name	The conditions for the electrospinning process	Applications in tissue engineering	References
Natural	Collagen	Collagen (type I), in the presence of EDC and NHS (as chemical crosslinking) solvent: ethanol-PBS 5 mL syringe (21 gauge), voltage: 20 kV, pump rate: 0.5 ml/h, relatively low humidity (20%), the speed of drum rotating: 5 m s^{-1}, and with a distance of 12 cm from the needle and the electrospun fibers Results: • The mean diameter of obtained fibers: 0.42 ± 0.11 μm • – The increase of the viscosity by adding crosslinker EDC to the collagen solution	To produce water-insoluble nano-fibers and create a uniaxial tensile behavior similar to native tissue.	[66]
		The syringe with an 18-gauge, rate of source solution by syringe pump: 0 to 25 ml/h, solvent: HFP, voltage: 15–30 kV, mandrel rotated: ~500 rpm, the optimal air gap distance: ~125 mm Results: • 100 nm fibers with the 67 nm banding pattern • The control of fiber orientation led to tailoring subtle mechanical properties into the matrix • The promotion of cell growth and its penetration into the electrospun collagen matrix	To optimize the concentrations, input voltage, delivery rate, air gap distance, and mandrel motion in electrospinning of collagen (type-I)	[67]
		Solvent: hexafluoro-2-propanol (HFP), syringe pump, and a 3 mL syringe with a 27½-G needle, voltage: 15 kV, distance between the needle tip and the collector: 15 cm Results: • The improvement of adhesion and proliferation of rabbit conjunctiva fibroblast on aligned collagen scaffolds	Blood vessel tissue engineering	[68]
	Gelatin	Solvent: HFP, a syringe with a needle diameter of 500 lm, injection: onto a metal collector, injection rate: 0.06 ml/h, Voltage: 12 kV/8 cm, temperature: 37°C. Results: • The co-solvents can be effective for electrospinning water-soluble natural polymers	Tissue engineering	[69]
	Fibrinogen	Solvent: HFP, 5.0-mL syringe (18-gauge needle) into a syringe pump, infection rate: 1.8 mL/h, voltage:	To design biomimicking fibrous scaffolds for tissue	[70]

The type of polymer	Name	The conditions for the electrospinning process	Applications in tissue engineering	References
		22 kV, distance between the needle tip and grounded target: 10 cm, rotating rate: 500 rpm. Results: • The efficient penetration of rat cardiac fibroblasts an electrospun matrix • The excellent candidate for soft tissue applications	engineering applications	
		4 mm diameter syringe with a 23¾-gauge, the syringe pump flow rate: 1.9 ml/h, voltage: 22 kV, distance between the needle tip and collector: 12.5 cm or 20 cm. Results: • High extensibility, low modulus	To apply in tissue engineering and polymer composite and biomimetic.	[71]
	Hyaluronic acid	Voltage: 0–40 kV, solvent: HCL, the aluminum foil as the collecting plate, and distance between the two electrodes: 9.5 cm. Results: • The diameter of fibers was 49–74 nm	Tissue engineering	[72]
	Silk	The electrospinning process consists of a steel capillary tube (1.5 mm), volume flow rate: constant via the electric potential, the control of the distance between the capillary tip and collection (aluminum foil), and flow rate for creating a stable jet without dripping. Results: • The increase of vascular cell growth • The formation of a short cord-like structure derived from HAECs after 4 days. • The formation of interconnection network of capillary tubes with lumens, after 7 days. • Suitable mechanical properties along with slow degradability	Vascular/blood vessels tissue engineering	[73]
Synthetic	Nano/micro poly(L-lactic acid) fibers	Solvent: DMF and DCM, needle: 18-G and inner diameter of 1.2 mm, the distance between the needle tip and the collector: 10 cm, a rotating disk along with a flat aluminum plate, and rotating rate: 1000 rpm, voltage: 12 kV. Results: • Fiber diameter: 300–3500 nm • Suitable for the neural stem cells culture	Neural tissue engineering	[74]

The type of polymer	Name	The conditions for the electrospinning process	Applications in tissue engineering	References
	Poly(lactide-co-glycolide) (PLGA) electrospun matrix	Solvent: HFP, and voltage of 25 kV, the needle with18 gage, flow rate: 3.0 mL/h. Distance between the needle tip and circular mandrel: 15 cm, and rotation rate: 500 rpm. Results: • Suitable tissue composition and ideal mechanical properties. • Biocompatible with cell interactions and without systemic/local toxic effect.	Blood vessel tissue engineering	[75]
	Poly (caprolactone) nanofiber	Solvent: chloroform/DMF, a syringe pump with a needle (inner diameter: 0.21 mm). Distance between the needle tip and aluminum plate: 15 cm, rotating rate of aluminum: 1000 rpm, voltage: 15 kV. Results: • Improvement of endothelial cell spreading. • The promotion of proliferation and control and growth of cell orientation.	Blood vessel tissue engineering	[76]
Copolymer/ hybrid	PCL/HA hybrid scaffold	Volume flow rate: 1.5 ml/h with a needle (ID = 0.8 mm), in the range of voltage: 15–20 kV, the distance between the capillary tube and the grounded target: 12–15 cm, and rotating rate: 300 r/min. Results: • The improvement of mechanical properties • The improvement of biochemical functions of nanofibers • The increase in the motility of skin fibroblast • The decrease in the limitations of electrospun scaffolds	To fabricate skin tissue-engineered scaffold	[77]
	HA–collagen nanofibrous matrix	Solvent: formic acid: HFIP (30/ 70 v/v), 5 ml syringe with a needle tip diameter of 0.96 mm, solution flow rate: 2.54 ml h^{-1}, voltage: 23 kV, the distance between needle tip and collector: 10 cm. Results: • Mean diameter of fibers: 200 nm • The structure similar to native ECM • As wound dressing for scarless skin regeneration	To fabricate a substrate for skin regeneration	[78]
	Sodium alginate/poly (vinyl alcohol)	Voltage: 17 kV, syringe pump with 10 ml syringe and needle of 21 gauges, the target: aluminum foil	To produce antibacterial wound dressings	[79]

The type of polymer	Name	The conditions for the electrospinning process	Applications in tissue engineering	References
	nanofibrous containing nano ZnO	fixed on a wooden stand, spinning rate: 0.1 ml/h, the distance between the needle tip and aluminum target: 5 cm. Results: • The increase in thermal stability of sodium alginate/PVA fibers, due to the existence of ZnO		
	Alginate-PEO nanofibers containing lavender essential oil	Solvent: water and DMF, syringe with a 23-gauge needle, syringe pump flow rate: 0.5 mL/h, voltage: 25 kV, distance between the needle tip and aluminum collector: 20 cm). Results: • Reduction of the production of pro-inflammatory cytokines • Wound healing and skin regeneration	To design a hybrid scaffold with antibacterial and anti-inflammatory activities	[80]
	Composite nanofibers of chitin/silk fibroin	Solvent: 1,1,1,3,3,3-hexafluoro-2-propanol (HFIP), distance between the needle tip and target: 7 cm, voltage: 17 kV, the mass flow rate: 4 ml/h, room temperature. Results: • The gradual decrease in the diameter of fibers occurred by increased chitin concentration • The improvement of cell attachment, spreading of epidermal keratinocyte and dermal fibroblast	Skin tissue engineering	[81]
	PLLA-gelatin scaffold with a layer of electrospun PDLA	Solvent: DCM and DMF For PDLA solution: plastic syringe with the 18-gauge needle, at the rate of 5 ml $h-^{1}$, rotating (~2000 rpm), and the distance: 15 cm, voltage: +12 kV and −5 kV. For gelatin/PLLA scaffold: the working distance: 5 cm (onto the PDLA layer), and voltages: +18 kV and −7 kV Results: • The increase of compressive mechanical properties. • Hydroxyapatite incorporation with PLLA/gelatin scaffold provided a significant effect on the mineral depositions on this scaffold.	Bone tissue engineering	[82]

Table 2.
The used polymers in the electrospinning process to produce nano/microfibers.

compared to other types of collagens (i.e., type II and type III) [85]. Given that collagen possesses a low young modulus (0.8 GPa) [83], the processes, such as chemical modification [by covalent of amine/imine linkage], cross-linking [by Glutaraldehyde (GA), NHS, and EDC, genipin as the cross-linking agent], and physical treatment [by UV irradiation, gamma radiation and dehydrothermal treatment (DHT)], can be led to an increase in mechanical properties of electrospun nanofibers based on this biopolymer [84, 86–96].

4.1.2 Gelatin

Gelatin derived from collagen is also one of the other biopolymers whose surface charge depends on the gelatin processing methods, such as acidic/alkaline processing [97]. The mechanical strength of this biopolymer can also increase via the physical mixing with other polymers or cross-linking process and immersing of gelatin-based scaffolds into glutaraldehyde (25%), carbodiimides, and genipin [83, 98].

4.1.3 Chitosan

Chitosan is another biopolymer that is widely used in biomedical/tissue engineering applications due to its low toxicity, non-immunogenic, biodegradability, and antibacterial properties [51, 98]. This polysaccharide as a cationic biopolymer can interact with structural molecules of the ECM due to having positively charged and be led to the formation of two-component scaffolds with suitable physicomechanical/biological properties when mixed with other anionic biopolymers [99].

4.1.4 Fibrinogen

Fibrinogen is a soluble biopolymer derived from the blood plasma and plays an important role in tissue engineering applications and the development of electrospun substrates/scaffolds [100, 101]. Based on the reports, fibrinogen electrospun fibers are more extensible and elastic compared to other biopolymer-based electrospun fibers [98, 102]. Notably, the mechanical resistance of this biopolymer and its degradation rate can be controlled by crosslinking process or supplementing culture medium [103].

4.1.5 Elastin

Elastin is a biopolymer of highly insoluble with difficult processing, hence the approaches, such as cross-linking and blending with other polymers can be effective in reaching elastin-based scaffolds [104].

4.1.6 Silk fibroin

Silk is a protein-biopolymer derived from Bombyx more (most commonly)/other insects and included two proteins of fibroin (70–80%) and sericin (20–30%) or other insects [105–109]. This biomacromolecular possesses suitable biocompatibility, biodegradability, and mechanical properties and plays an important role in the biomedical applications and development of engineered scaffolds [110–114]. Based on the reports, the electrospun nanofibers of this biopolymer can modulate cellular interactions (such as adhesion, spread, the expression level of genes and proteins) [115, 116].

4.1.7 Hyaluronic acid

Hyaluronic acid (HA) is known as a linear polysaccharide and plays an important role in the ECM structure [117, 118]. This glycosaminoglycan (GAG) is a suitable candidate for hydrogel production and tissue regeneration due to its molecular weight (100–8000 kDa) and hygroscopic nature [118–120]. However, the high viscosity of produced hydrogels leads to limitations in the electrospinning process, hence, the electrospun fibers can form by blending/dissolving this hydrogel with other polymers/in other solvents [65].

4.1.8 Alginate

Alginate is a copolymer derived from β-D-mannuronic acid and α-L-guluronic acid that is widely used in tissue engineering applications [121, 122]. This biopolymer is known as one of the other polysaccharides that are commercially produced from brown algae or some bacteria, such as *Azotobacter chroococcum*, *Azotobacter vinelandii*, and some species of *Pseudomonas* [121, 123]. The salt of this polymer (alginate sodium) is water-soluble and can also form a highly viscous solution at very low concentrations [124]. Moreover, the electrospinning of alginate sodium solutions carries out in presence of synthetic polymers (PVA, PEO, etc.), owing to the reduction of the repulsive forces among the poly-anionic chains of alginate. Methods, such as cross-linking, are also necessary for resulting scaffold stability in the aqueous environment [65].

4.2 Synthetic polymers

4.2.1 Polylactic acid (PLA)

PLA as thermoplastic polymer fabricates from the polymerization of lactic acid and possesses isomers of poly(L-lactic acid) and poly(D-lactic acid) [125, 126]. This aliphatic polyester possesses biodegradability properties and plays a critical role in the design of tissue-engineered scaffolds [127]. However, PLA is usually copolymerized with other polymers or made as a composite due to its low modulus, especially in bone tissue engineering [128, 129].

4.2.2 Polycaprolactone (PCL)

This polymer is a semicrystalline polyester that is widely used in the applications of tissue engineering due to its biodegradability, nontoxicity, and biocompatibility properties, as well as mechanical strength [130–132]. This synthetic polymer can improve cellular penetration into the engineered scaffolds due to the presence of cell recognition sites [51, 133, 134]. However, the degraded product of this polymer (the acids of polylactide and glycolide) affects on stability and functions of proteins and bioactive molecules. PCL is also known as the most famous synthetic polymer in the design of bone-engineered scaffolds, due to its low degradation rate and high modulus [83].

4.2.3 Polyglycolic acid (PGA)

PGA is the aliphatic thermoplastic polyester (simple linear) that is usually applied in the applications of bone tissue engineering. This polymer possesses a high young modulus (7 GPa) and can be completely degraded within 4–6 months [133, 135].

4.2.4 Polyethylene glycol (PEG)

PEG is known as one of the most popular synthetic polymers in tissue engineering applications and can lead to a promotion of cellular adhesion and improvement of cell–cell signaling, due to its hydrophilic properties and interactions with the chains of polysaccharides or peptides [51, 136]. Notably, copolymerization of this polymer with other hydrophobic polymers, such as polyglycolic acid, polycaprolactone, and polylactic acid, can lead to an increase in the degradation rate of these polymers and neutralization of their acidic products [137, 138].

4.3 Copolymer/hybrid polymers

Tissue engineering is a strategy for the design of tissue-liked scaffolds/ substrates (in terms of biological and physiological functions). In this field, the combination of natural or synthetic polymers as copolymers or hybrid scaffolds can be played an important role in overcoming the limitations of mono-component systems and improving the interactions of cell/cell and cell/scaffold. Sodium alginate/PVA electrospun mats are a sample from these copolymers that aimed to prepare the antibacterial substrates [79]. Such substrates can be used in wound dressings to reduce wound infection and prevent scars [79, 139]. PCL/HA nanofibrous scaffolds are also one of the other composite scaffolds that can provide substrates with better mechanical and biochemical properties. Based on the reports, such scaffolds lead to an increase in fibroblasts infiltration into the scaffold, cellular proliferation, and consequently tissue regeneration [77]. In this field, the combination of HA and collagen has been reported as an ideal matrix in electrospun dressings that play an important role in reducing scar via proteinase secretion and metalloproteinase inhibition [78]. Alginate-PEO nanofibers containing lavender essential oil are another scaffold that can be used for wound healing and reducing the production of pro-inflammatory cytokines [80]. The PLLA-gelatin scaffold with a layer of electrospun PDLA is one of the other samples in this field [82].

5. Application of electrospun polymers in tissue engineering

The cells play an important role in the formation of organ-dependent extracellular matrix (ECM), and to this end, they need microenvironments to improve their functions. However, the body is unable to repair damaged tissue when tissue damage is severe or extensive. In this field, although, the xenografts, autografts, and allografts approaches can be used, however, the problems of donor sites, antigenicity, and immunogenicity have been limited to the use of these therapeutic methods [51]. In recent years, tissue engineering as a new method has been overcome the mentioned problems via designing engineered substrates with suitable physicomechanical and biological properties [3, 51]. There are many studies that show electrospun substrates can be effective in this field. Such that, the electrospinning process of natural and synthetic polymers can help to address cell requirements, improve its functions, and finally regenerate damaged tissue.

Blood vessels are one of the important organs in the body that need to be repaired quickly in case of damage. Vascular-engineered scaffolds play a key role in this regard so that electrospun matrices based on the natural/synthetic/hybrid polymers can support the adhesion, differentiation, and proliferation of vascular cells and be led to tissue regeneration [140]. Accordingly, the study of Bondar et al. shows that there is ideal intercellular contact between endothelial cells on the nano/micro-scale electrospun silk fiber that led to vascular endothelial cadherin expression [141]. Soffer et al. also designed silk fibroin into a tubular structure (inner-diameter: ~3 mm, the average wall thickness: 0.15 mm). They reported that the average strength of such scaffolds is much more than collagen scaffolds [142].

Skin is another organ that possesses an important role in the protection of the body against infection and environmental agents. Such that, loss of a large surface of the skin due to burn, wound, etc. can lead to patient death. However, dermis-engineered scaffolds can be an ideal therapeutic option for skin regeneration [6]. In this field, Min et al. stated that silk electrospun substrates with coating collagen I can increase adhesion and spreading of keratinocytes. Moreover, they found that laminin coating can stimulate cellular spreading, but not cellular adhesion [112]. Based on the studies of Pezeshki-Modaress et al., the gelatin/chitosan electrospun scaffolds as the structures containing protein and polysaccharide play a crucial role in the proliferation of dermal fibroblast cells and wound healing. Such scaffolds can maintain their morphology in culture medium and increase the proliferation and attachment of cells [143]. Law et al. also assessed electrospun collagen nanofibers for applications of skin tissue engineering. They stated that collagen nanofibers possess low mechanical properties, hence are usually cross-linked/blended with synthetic polymers [144]. In this field, Jin et al. showed that collagen/poly(l-lactic acid)–co-poly(3-caprolactone) electrospun scaffolds can differentiate mesenchymal stem cells to epidermal and lead to an increase in cell proliferation [145].

One of the other applications of electrospun scaffolds is related to the design of calcified extracellular matrices. These substrates consist of calcium phosphates, carbonated hydroxyapatite, growth factors, and bone marrow stromal cells (MSCs) or osteoblasts and osteoblast-like cells, such as osteoprogenitors and osteocytes [140]. In this regard, Ki et al. designed the 3D silk matrices to produce bone with MC3T3-E1 cells. They found that cell proliferation and spreading increase on the 3D matrices compared to 2D matrices. This can be due to the higher porosity of 3D matrices that provides better cellular adhesion [146]. In another study, Yin et al. showed that electrospun scaffolds can be effective on the regeneration of various tissues, via topography-dependent induction of lineage-specific differentiation [147]. They evaluated the differentiation pathway of MSCs in forming new tissues and observed that these scaffolds can lead to the formation of tendon-like tissue in the Achilles tendon injury models, as well as, chondrogenesis and bone tissue formation via ossification. Delgado-Rangel et al. also developed collagen/poly (vinyl alcohol)/ chondroitin sulfate and collagen/poly (vinyl alcohol)/hyaluronic acid 3D electrospun scaffolds to apply tissue engineering [148]. Based on their reports, these scaffolds can increase the biocompatible cross-linker. Moreover, the scaffolds possess the behavior of pH-sensitive swelling and can be used in drug delivery systems.

Flaig et al. also studied the application of electrospun scaffolds in cardiac tissue engineering. They found that electrospun scaffolds based on poly (glycerol sebacate) elastomer and poly (lactic acid) can induce neovascularization without the inflammatory responses and support cardiomyocyte development [149]. Moreover, Vogt et al. stated that poly (ε-caprolactone)/poly (glycerol sebacate)-based electrospun scaffolds possess better mechanical properties compared to native myocardium; hence can be potentially suitable to apply cardiac tissue engineering [150].

Collectively, the studies indicate that electrospun scaffolds possess an inductive role in the regeneration of tissues and the use of hybrid polymers can provide effective insight into the design and development of smart scaffolds for applications of tissue engineering.

6. Challenges and resolutions of the electrospinning process

The techniques of tissue engineering hold promise for developing functional networks similar to native tissue. The designed substrates in this technique can support the formation of 3D tissues by mimicking ECM functions. Among the fabrication techniques of the engineered scaffold, the electrospinning process is known as an outstanding one that can produce a nonwoven structure.

Although this method is considered as the potential technique in the design of tissue-engineered scaffolds, however, it possesses limitations of mechanical strength and cellular infiltration in the application of load-bearing.

Based on the reports, the optimum size of pores for cell infiltration to tissue is in a range of 100–500 μm [151], while, the size of pores in electrospun scaffolds is much lower than the mentioned size. This can lead to inhibition of vascular growth and the creation of the hypoxic region. Moreover, the density of the fibers in electrospun substrates can be one of the reasons for poor cellular infiltration [152].

There are several solutions to overcome these limitations that pore architecture of scaffolds and their surface morphology control are some of the most important solutions. Generally, the diameter of fibers strongly relates to the pore diameter in the electrospun substrates, so that, fibers with smaller diameters lead to smaller pores. Hence, attention to surface topography plays a crucial role in the removal of waste and diffusion of nutrients [153, 154].

Indeed, the manipulation of characteristics of the electrospun scaffolds for enlarging the diameter of pores or reduction of the fibers density, help migration of cells into internal parts of the scaffold. There are other four approaches in this field, including [155]:

1. Adding biological factors

2. Electrospraying or layering of cells

3. Dynamic cellular culture

4. Combination of electrospinning with other fibers-fabrication techniques.

7. Conclusion

The electrospinning process is known as the powerful, simple, and inexpensive tool to fabricate tissue-engineering substrates that are capable of the formation of ECM-mimicking networks. Although, this technique, in clinical applications; has the limitations, such as low cellular infiltration, high-density of fibers, possible toxicity of solvent/cross-linker, and insufficient mechanical strength. However, some solutions, such as the increased diameter of pores, reduced density of fibers, and electrospinning polymers along with cells can overcome these problems. Combining robust materials or structures also provides the more robust electrospun substrate for the design and production of tissue substitutes with the desired target.

Author details

Azadeh Izadyari Aghmiuni[1*], Arezoo Ghadi[2], Elmira Azmoun[3], Niloufar Kalantari[2], Iman Mohammadi[2] and Hossein Hemati Kordmahaleh[2]

1 Department of Nanobiotechnology, Pasteur Institute of Iran, Tehran, Iran

2 Department of Chemical Engineering, Ayatollah Amoli Branch, Islamic Azad University, Amol, Iran

3 Department of Chemical Engineering, Islamic Azad University, Gaemshahr branch, Mazandaran, Iran

*Address all correspondence to: azadeh.izadyari@gmail.com

References

[1] Jun I, Han HS, Edwards JR, Jeon H. Electrospun fibrous scaffolds for tissue engineering: Viewpoints on architecture and fabrication. International Journal of Molecular Sciences. 2018;**19**. DOI: 10.3390/ijms19030745

[2] Chen F-M, Liu X. Advancing biomaterials of human origin for tissue engineering. Progress in Polymer Science. 2016;**53**:86-168

[3] Izadyari Aghmiuni A, Heidari Keshel S, Sefat F, AkbarzadehKhiyavi A. Fabrication of 3D hybrid scaffold by combination technique of electrospinning-like and freeze-drying to create mechanotransduction signals and mimic extracellular matrix function of skin. Materials Science and Engineering: C. 2021;**120**:111752

[4] Thomas K, Engler AJ, Meyer GA. Extracellular matrix regulation in the muscle satellite cell niche. Connective Tissue Research. 2015;**56**:1-8

[5] Wen S-L, Feng S, Tang S-H, Gao J-H, Zhang L, Tong H, et al. Collapsed reticular network and its possible mechanism during the initiation and/or progression of hepatic fibrosis. Scientific Reports. 2016;**6**:1-11

[6] Izadyari Aghmiuni A, Heidari Keshel S, Sefat F, Akbarzadeh Khiyavi A. Quince seed mucilage-based scaffold as a smart biological substrate to mimic mechanobiological behavior of skin and promote fibroblasts proliferation and h-ASCs differentiation into keratinocytes. International Journal of Biological Macromolecules. 2020;**142**:668-679

[7] Caralt M, Uzarski JS, Iacob S, Obergfell KP, Berg N, Bijonowski BM, et al. Optimization and critical evaluation of decellularization strategies to develop renal extracellular matrix scaffolds as biological templates for organ engineering and transplantation. American Journal of Transplantation. 2015;**15**:64-75

[8] Bera B. Literature review on electrospinning process (a fascinating fiber fabrication technique). International Journal of Interdisciplinary Research (IJIR). 2016;**2**:972-984

[9] Pillay V, Dott C, Choonara YE, Tyagi C, Tomar L, Kumar P, et al. A review of the effect of processing variables on the fabrication of electrospun nanofibers for drug delivery applications. Journal of Nanomaterials. 2013;**2013**:1-22

[10] Mirjalili M, Zohoori S. Review for application of electrospinning and electrospun nanofibers technology in textile industry. Journal of Nanostructure in Chemistry. 2016;**6**:207-213

[11] Zafar M, Najeeb S, Khurshid Z, Vazirzadeh M, Zohaib S, Najeeb B, et al. Potential of electrospun nanofibers for biomedical and dental applications. Materials (Basel). 2016;**9**:73

[12] Correia DM, Ribeiro C, Ferreira JCC, Botelho G, Ribelles JLG, Lanceros-Méndez S, et al. Influence of electrospinning parameters on poly (hydroxybutyrate) electrospun membranes fiber size and distribution. Polymer Engineering and Science. 2014;**54**:1608-1617

[13] Motamedi AS, Mirzadeh H, Hajiesmaeilbaigi F, Bagheri-Khoulenjani S, Shokrgozar M. Effect of electrospinning parameters on morphological properties of PVDF nanofibrous scaffolds. Progress in Biomaterials. 2017;**6**:113-123

[14] Vatankhah E, Prabhakaran MP, Semnani D, Razavi S, Morshed M, Ramakrishna S. Electrospun tecophilic/gelatin nanofibers with potential for small diameter blood vessel tissue engineering. Biopolymers. 2014;**101**:1165-1180

[15] Krishnan KV, Columbus S, Krishnan LK. Alteration of electrospun scaffold properties by silver nanoparticle incorporation: Evaluation for blood vessel tissue engineering. Tissue Engineering Part A. 2015;**21**:S239-S240

[16] Sundaramurthi D, Krishnan UM, Sethuraman S. Electrospun nanofibers as scaffolds for skin tissue engineering. Polymer Reviews. 2014;**54**:348-376

[17] Chen S, Liu B, Carlson MA, Gombart AF, Reilly DA, Xie J. Recent advances in electrospun nanofibers for wound healing. Nanomedicine. 2017;**12**:1335-1352

[18] Norouzi M, Boroujeni SM, Omidvarkordshouli N, Soleimani M. Advances in skin regeneration: application of electrospun scaffolds. Advanced Healthcare Materials. 2015;**4**:1114-1133

[19] Wang C, Wang M. Electrospun multicomponent and multifunctional nanofibrous bone tissue engineering scaffolds. Journal of Materials Chemistry. 2017;**5**:1388-1399

[20] Xu T, Miszuk JM, Zhao Y, Sun H, Fong H. Electrospun polycaprolactone 3D nanofibrous scaffold with interconnected and hierarchically structured pores for bone tissue engineering. Advanced Healthcare Materials. 2015;**4**:2238-2246

[21] Kheradmandi M, Vasheghani-Farahani E, Ghiaseddin A, Ganji F. Skeletal muscle regeneration via engineered tissue culture over electrospun nanofibrous chitosan/PVA scaffold. Journal of Biomedical Materials Research Part A. 2016;**104**:1720-1727

[22] Elsayed Y, Lekakou C, Labeed F, Tomlins P. Smooth muscle tissue engineering in crosslinked electrospun gelatin scaffolds. Journal of Biomedical Materials Research Part A. 2016;**104**:313-321

[23] Liu W, Li Z, Zheng L, Zhang X, Liu P, Yang T, et al. Electrospun fibrous silk fibroin/poly (L-lactic acid) scaffold for cartilage tissue engineering. Tissue Engineering and Regenerative Medicine. 2016;**13**:516-526

[24] Fang J, Wang X, Lin T. Functional applications of electrospun nanofibers. Nanofibers-Production, Properties and Functional Applications. 2011;**14**:287-302

[25] Awang N, Ismail AF, Jaafar J, Matsuura T, Junoh H, Othman MHD, et al. Functionalization of polymeric materials as a high performance membrane for direct methanol fuel cell: A review. Reactive and Functional Polymers. 2015;**86**:248-258

[26] Khadka DB, Haynie DT. Protein- and peptide-based electrospun nanofibers in medical biomaterials. Nanomedicine Nanotechnology Biologie et Médecine. 2012;**8**:1242-1262

[27] Deitzel JK, Harris D, Tan NB. The effect of processing variables on the morphology of electrospun nanofibers and textiles. Polymer (Guildf). 2001;**42**:261-272

[28] Eda G, Shivkumar S. Bead-to-fiber transition in electrospun polystyrene. Journal of Applied Polymer Science. 2007;**106**:475-487

[29] Lee KH, Kim HY, Bang HJ, Jung YH, Lee SG. The change of bead morphology formed on electrospun polystyrene fibers. Polymer (Guildf). 2003;**44**: 4029-4034

[30] Yang Q, Li Z, Hong Y, Zhao Y, Qiu S, Wang C, et al. Influence of solvents on the formation of ultrathin uniform poly (vinyl pyrrolidone) nanofibers with electrospinning. Journal of Polymer Science Part B Polymer Physics. 2004; **42**:3721-3726

[31] McKee MG, Layman JM, Cashion MP, Long TE. Phospholipid nonwoven electrospun membranes. Science. 2006; **311**(80):353-355

[32] Koski A, Yim K, Shivkumar S. Effect of molecular weight on fibrous PVA produced by electrospinning. Materials Letters. 2004;**58**:493-497

[33] Zhao YY, Yang QB, Lu X-F, Wang C, Wei Y. Study on correlation of morphology of electrospun products of polyacrylamide with ultrahigh molecular weight. Journal of Polymer Science Part B Polymer Physics. 2005;**43**: 2190-2195

[34] Larrondo L, John Manley RS. Electrostatic fiber spinning from polymer melts. I. Experimental observations on fiber formation and properties. Journal of Polymer Science Part B Polymer Physics. 1981;**19**:909-920

[35] Sukigara S, Gandhi M, Ayutsede J, Micklus M, Ko F. Regeneration of Bombyx mori silk by electrospinning—part 1: processing parameters and geometric properties. Polymer (Guildf). 2003;**44**:5721-5727

[36] Zhang Y, Ouyang H, Lim CT, Ramakrishna S, Huang Z-M. Electrospinning of gelatin fibers and gelatin/PCL composite fibrous scaffolds. Journal of Biomedical Materials Research. 2005;**72B**:156-165

[37] Ki CS, Baek DH, Gang KD, Lee KH, Um IC, Park YH. Characterization of gelatin nanofiber prepared from gelatin–formic acid solution. Polymer (Guildf). 2005;**46**:5094-5102

[38] Haghi AK, Akbari M. Trends in electrospinning of natural nanofibers. Physica Status Solidi. 2007;**204**: 1830-1834

[39] Pham QP, Sharma U, Mikos AG. Electrospun poly(ε-caprolactone) microfiber and multilayer nanofiber/ microfiber scaffolds: Characterization of scaffolds and measurement of cellular infiltration. Biomacromolecules. 2006;**7**: 2796-2805

[40] Zhang C, Yuan X, Wu L, Han Y, Sheng J. Study on morphology of electrospun poly(vinyl alcohol) mats. European Polymer Journal. 2005;**41**:423-432

[41] Zong X, Kim K, Fang D, Ran S, Hsiao BS, Chu B. Structure and process relationship of electrospun bioabsorbable nanofiber membranes. Polymer (Guildf). 2002;**43**:4403-4412

[42] Huang C, Chen S, Lai C, Reneker DH, Qiu H, Ye Y, et al. Electrospun polymer nanofibres with small diameters. Nanotechnology. 2006;**17**: 1558-1563

[43] Yuan X, Zhang Y, Dong C, Sheng J. Morphology of ultrafine polysulfone fibers prepared by electrospinning. Polymer International. 2004;**53**: 1704-1710

[44] Buchko CJ, Chen LC, Shen Y, Martin DC. Processing and microstructural characterization of porous biocompatible

protein polymer thin films. Polymer (Guildf). 1999;**40**:7397-7407

[45] Li Z, Wang C. One-Dimensional Nanostructures: Electrospinning Technique and Unique Nanofibers. New York, Dordrecht, London: Springer Berlin Heidelberg; 2013. pp. 15-29. DOI: 10.1007/978-3-642-36427-3_2

[46] Wang X, Um IC, Fang D, Okamoto A, Hsiao BS, Chu B. Formation of water-resistant hyaluronic acid nanofibers by blowing-assisted electro-spinning and non-toxic post treatments. Polymer (Guildf). 2005;**46**:4853-4867

[47] Li D, Wang Y, Xia Y. Electrospinning nanofibers as uniaxially aligned arrays and layer-by-layer stacked films. Advanced Materials. 2004;**16**: 361-366

[48] Xu C. Aligned biodegradable nanofibrous structure: A potential scaffold for blood vessel engineering. Biomaterials. 2004;**25**:877-886

[49] Ki CS, Kim JW, Hyun JH, Lee KH, Hattori M, Rah DK, et al. Electrospun three-dimensional silk fibroin nanofibrous scaffold. Journal of Applied Polymer Science. 2007;**106**:3922-3928

[50] Mit-uppatham C, Nithitanakul M, Supaphol P. Ultrafine electrospun polyamide-6 fibers: Effect of solution conditions on morphology and average fiber diameter. Macromolecular Chemistry and Physics. 2004;**205**: 2327-2338

[51] Aghmiuni AI, Baei MS, Keshel SH, Khiyavi AA. Design of novel 3D-scaffold as a potential material to induct epidermal-dermal keratinocytes of human-adipose-derived stem cells and promote fibroblast cells proliferation for skin regeneration. Fibers and Polymers. 2020;**21**:33-44

[52] Biazar E, Keshel SH. The healing effect of stem cells loaded in nanofibrous scaffolds on full thickness skin defects. Journal of Biomedical Nanotechnology. 2013;**9**:1471-1482

[53] Bassas-Galia M, Follonier S, Pusnik M, Zinn M. Natural polymers. In: Bioresorbable Polymers for Biomedical Applications. Amsterdam, Netherlands: Elsevier; 2017. pp. 31-64. DOI: 10.1016/B978-0-08-100262-9.00002-1

[54] Wang S, Lu L, Wang C, Gao C, Wang X. International Journal of Polymer Science. 2010;**2010**:1-2

[55] Sell SA, Wolfe PS, Garg K, McCool JM, Rodriguez IA, Bowlin GL. The use of natural polymers in tissue engineering: A focus on electrospun extracellular matrix analogues. Polymers (Basel). 2010;**2**: 522-553

[56] Place ES, George JH, Williams CK, Stevens MM. Synthetic polymer scaffolds for tissue engineering. Chemical Society Reviews. 2009;**38**: 1139-1151

[57] Kitsara M, Agbulut O, Kontziampasis D, Chen Y, Menasché P. Fibers for hearts: A critical review on electrospinning for cardiac tissue engineering. Acta Biomaterialia. 2017;**48**: 20-40

[58] Sengupta P, Prasad BLV. Surface modification of polymeric scaffolds for tissue engineering applications. Regenerative Engineering and Translational Medicine. 2018;**4**:75-91

[59] Ahire JJ, Robertson D, Neveling DP, Van Reenen AJ, Dicks LMT. Hyaluronic acid-coated poly(d,l-lactide) (PDLLA) nanofibers prepared by electrospinning and coating. RSC Advances. 2016;**6**: 34791-34796

[60] Bosworth LA, Alam N, Wong JK, Downes S. Investigation of 2D and 3D electrospun scaffolds intended for tendon repair. Journal of Materials Science: Materials in Medicine. 2013;**24**: 1605-1614

[61] Kumbar SG, Nukavarapu SP, James R, Nair LS, Laurencin CT. Electrospun poly(lactic acid-co-glycolic acid) scaffolds for skin tissue engineering. Biomaterials. 2008;**29**:4100-4107

[62] Mehrasa M, Asadollahi MA, Ghaedi K, Salehi H, Arpanaei A. Electrospun aligned PLGA and PLGA/gelatin nanofibers embedded with silica nanoparticles for tissue engineering. International Journal of Biological Macromolecules. 2015;**79**:687-695

[63] Chen SH, Chen CH, Shalumon KT, Chen JP. Preparation and characterization of antiadhesion barrier film from hyaluronic acid-grafted electrospun poly(caprolactone) nanofibrous membranes for prevention of flexor tendon postoperative peritendinous adhesion. International Journal of Nanomedicine. 2014;**9**: 4079-4092

[64] Nanofibers and their applications as templates for nanofibers preparation metal oxide nanofibers preparation

[65] Mele E. Electrospinning of natural polymers for advanced wound care: Towards responsive and adaptive dressings. Journal of Materials Chemistry B. 2016;**4**:4801-4812

[66] Meng L, Arnoult O, Smith M, Wnek GE. Electrospinning of in situ crosslinked collagen nanofibers. Journal of Materials Chemistry. 2012;**22**: 19412-19417

[67] Matthews JA, Wnek GE, Simpson DG, Bowlin GL. Electrospinning of collagen nanofibers. Biomacromolecules. 2002;**3**:232-238

[68] Zhong S, Teo WE, Zhu X, Beuerman RW, Ramakrishna S, Yung LYL. An aligned nanofibrous collagen scaffold by electrospinning and its effects on in vitro fibroblast culture. The Journal of Biomedical Materials Research Part A An Official Journal of Society for Biomaterials Australian Society for Biomaterials; Korean Society for Biomaterials. 2006;**79**:456-463

[69] Song J-H, Kim H-E, Kim H-W. Production of electrospun gelatin nanofiber by water-based co-solvent approach. Journal of Materials Science: Materials in Medicine. 2008;**19**:95-102

[70] McManus MC, Boland ED, Simpson DG, Barnes CP, Bowlin GL. Electrospun fibrinogen: Feasibility as a tissue engineering scaffold in a rat cell culture model. The Journal of Biomedical Materials Research Part A An Official Journal of Society for Biomaterials Australian Society for Biomaterials.; Korean Society for Biomaterials. 2007; **81**:299-309

[71] Carlisle CR, Coulais C, Namboothiry M, Carroll DL, Hantgan RR, Guthold M. The mechanical properties of individual, electrospun fibrinogen fibers. Biomaterials. 2009;**30**:1205-1213

[72] Um IC, Fang D, Hsiao BS, Okamoto A, Chu B. Electro-spinning and electro-blowing of hyaluronic acid. Biomacromolecules. 2004;**5**:1428-1436

[73] Zhang X, Baughman CB, Kaplan DL. In vitro evaluation of electrospun silk fibroin scaffolds for vascular cell growth. Biomaterials. 2008;**29**:2217-2227

[74] Yang F, Murugan R, Wang S, Ramakrishna S. Electrospinning of nano/micro scale poly(L-lactic acid) aligned

fibers and their potential in neural tissue engineering. Biomaterials. 2005;**26**: 2603-2610

[75] Stitzel J, Liu J, Lee SJ, Komura M, Berry J, Soker S, et al. Controlled fabrication of a biological vascular substitute. Biomaterials. 2006;**27**:1088-1094

[76] Ma Z, He W, Yong T, Ramakrishna S. Grafting of gelatin on electrospun poly (caprolactone) nanofibers to improve endothelial cell spreading and proliferation and to control cell orientation. Tissue Engineering. 2005;**11**: 1149-1158

[77] Qian Y, Li L, Jiang C, Xu W, Lv Y, Zhong L, et al. The effect of hyaluronan on the motility of skin dermal fibroblasts in nanofibrous scaffolds. International Journal of Biological Macromolecules. 2015;**79**:133-143

[78] Hsu F-Y, Hung Y-S, Liou H-M, Shen C-H. Electrospun hyaluronate–collagen nanofibrous matrix and the effects of varying the concentration of hyaluronate on the characteristics of foreskin fibroblast cells. Acta Biomaterialia. 2010; **6**:2140-2147

[79] Shalumon KT, Anulekha KH, Nair SV, Nair SV, Chennazhi KP, Jayakumar R. Sodium alginate/poly (vinyl alcohol)/ nano ZnO composite nanofibers for antibacterial wound dressings. International Journal of Biological Macromolecules. 2011;**49**:247-254

[80] Hajiali H, Summa M, Russo D, Armirotti A, Brunetti V, Bertorelli R, et al. Alginate–lavender nanofibers with antibacterial and anti-inflammatory activity to effectively promote burn healing. Journal of Materials Chemistry B. 2016;**4**:1686-1695

[81] Park KE, Jung SY, Lee SJ, Min BM, Park WH. Biomimetic nanofibrous scaffolds: Preparation and characterization of chitin/silk fibroin blend nanofibers. International Journal of Biological Macromolecules. 2006;**38**: 165-173

[82] Andric T, Wright LD, Taylor BL, Freeman JW. Fabrication and characterization of three-dimensional electrospun scaffolds for bone tissue engineering. Journal of Biomedical Materials Research Part A. 2012;**100**: 2097-2105

[83] Bhattarai DP, Aguilar LE, Park CH, Kim CS. A review on properties of natural and synthetic based electrospun fibrous materials for bone tissue engineering. Membranes. 2018;**8**:62. DOI: 10.3390/membranes8030062

[84] Zuber M, Zia F, Zia KM, Tabasum S, Salman M, Sultan N. Collagen based polyurethanes—A review of recent advances and perspective. International Journal of Biological Macromolecules. 2015;**80**:366-374

[85] Shields KJ, Beckman MJ, Bowlin GL, Wayne JS. Mechanical properties and cellular proliferation of electrospun collagen type II. Tissue Engineering. 2004;**10**:1510-1517

[86] Boland ED, Matthews JA, Pawlowski KJ, Simpson DG, Wnek GE, Bowlin GL. Electrospinning collagen and elastin: preliminary vascular tissue engineering. Frontiers in Bioscience. 2004;**9**:e32

[87] Barnes CP, Pemble CW IV, Brand DD, Simpson DG, Bowlin GL. Cross-linking electrospun type II collagen tissue engineering scaffolds with carbodiimide in ethano. Tissue Engineering. 2007;**13**:1593-1605

[88] Sell SA, McClure MJ, Garg K, Wolfe PS, Bowlin GL. Electrospinning of collagen/biopolymers for regenerative

medicine and cardiovascular tissue engineering. Advanced Drug Delivery Reviews. 2009;**61**:1007-1019

[89] Liu S, Wu J, Liu X, Chen D, Bowlin GL, Cao L, et al. Osteochondral regeneration using an oriented nanofiber yarn-collagen type I/hyaluronate hybrid/TCP biphasic scaffold. Journal of Biomedical Materials Research Part A. 2015;**103**:581-592

[90] Wang W, Zhang Y, Ye R, Ni Y. Physical crosslinkings of edible collagen casing. International Journal of Biological Macromolecules. 2015;**81**:920-925

[91] Takitoh T, Bessho M, Hirose M, Ohgushi H, Mori H, Hara M. Gamma-cross-linked nonfibrillar collagen gel as a scaffold for osteogenic differentiation of mesenchymal stem cells. Journal of Bioscience and Bioengineering. 2015;**119**: 217-225

[92] Maslennikova A, Kochueva M, Ignatieva N, Vitkin A, Zakharkina O, Kamensky V, et al. Effects of gamma irradiation on collagen damage and remodeling. International Journal of Radiation Biology. 2015;**91**:240-247

[93] Suchý T, Šupová M, Sauerová P, Verdánová M, Sucharda Z, Rýglová Š, et al. The effects of different cross-linking conditions on collagen-based nanocomposite scaffolds—An in vitro evaluation using mesenchymal stem cells. Biomedical Materials. 2015;**10**: 65008

[94] Yoshioka SA, Goissis G. Thermal and spectrophotometric studies of new crosslinking method for collagen matrix with glutaraldehyde acetals. Journal of Materials Science: Materials in Medicine. 2008;**19**:1215-1223

[95] Vrana NE, Builles N, Kocak H, Gulay P, Justin V, Malbouyres M, et al. EDC/NHS cross-linked collagen foams as scaffolds for artificial corneal stroma. Journal of Biomaterials Science, Polymer Edition. 2007;**18**:1527-1545

[96] Liu T, Wang Z. Collagen crosslinking of porcine sclera using genipin. Acta Ophthalmologica. 2013;**91**: e253-e257

[97] Young S, Wong M, Tabata Y, Mikos AG. Gelatin as a delivery vehicle for the controlled release of bioactive molecules. Journal of Controlled Release;**109, 2005**: 256-274

[98] Nemati S, Kim S, Shin YM, Shin H. Current progress in application of polymeric nanofibers to tissue engineering. Nano Convergence. 2019. DOI: 10.1186/s40580-019-0209-y

[99] Nilsen-Nygaard J, Strand SP, Vårum KM, Draget KI, Nordgård CT. Chitosan: Gels and interfacial properties. Polymers (Basel). 2015;7:552-579

[100] Wnek GE, Carr ME, Simpson DG, Bowlin GL. Electrospinning of nanofiber fibrinogen structures. Bachelor of Engineering, Virginia Commonwealth University. 2003:1-4

[101] Rothwell SW, Sawyer E, Dorsey J, Flournoy WS, Settle T, Simpson D, et al. Wound healing and the immune response in swine treated with a hemostatic bandage composed of salmon thrombin and fibrinogen. Journal of Materials Science: Materials in Medicine. 2009;**20**:2155-2166

[102] Baker S, Sigley J, Helms CC, Stitzel J, Berry J, Bonin K, et al. The mechanical properties of dry, electrospun fibrinogen fibers. Materials Science and Engineering: C. 2012;**32**:215-221

[103] McManus MC, Boland ED, Koo HP, Barnes CP, Pawlowski KJ, Wnek GE, et al.

Mechanical properties of electrospun fibrinogen structures. Acta Biomaterialia. 2006;**2**:19-28

[104] Leach JB, Wolinsky JB, Stone PJ, Wong JY. Crosslinked α-elastin biomaterials: towards a processable elastin mimetic scaffold. Acta Biomaterialia. 2005;**1**:155-164

[105] Wang Y, Kim H-J, Vunjak-Novakovic G, Kaplan DL. Stem cell-based tissue engineering with silk biomaterials. Biomaterials. 2006;**27**: 6064-6082

[106] Qi C, Xu L, Deng Y, Wang G, Wang Z, Wang L. Sericin hydrogels promote skin wound healing with effective regeneration of hair follicles and sebaceous glands after complete loss of epidermis and dermis. Biomaterials Science. 2018;**6**:2859-2870

[107] Wray LS, Hu X, Gallego J, Georgakoudi I, Omenetto FG, Schmidt D, et al. Effect of processing on silk-based biomaterials: Reproducibility and biocompatibility. Journal of Biomedical Materials Research Part B: Applied Biomaterials. 2011;**99**:89-101

[108] Rockwood DN, Preda RC, Yücel T, Wang X, Lovett ML, Kaplan DL. Materials fabrication from Bombyx mori silk fibroin. Nature Protocols. 2011;**6**: 1612-1631

[109] Melke J, Midha S, Ghosh S, Ito K, Hofmann S. Silk fibroin as biomaterial for bone tissue engineering. Acta Biomaterialia. 2016;**31**:1-16

[110] Meinel L, Betz O, Fajardo R, Hofmann S, Nazarian A, Cory E, et al. Silk based biomaterials to heal critical sized femur defects. Bone. 2006;**39**:922-931

[111] Lin S, Chen M, Jiang H, Fan L, Sun B, Yu F, et al. Green electrospun grape seed extract-loaded silk fibroin nanofibrous mats with excellent cytocompatibility and antioxidant effect. Colloids Surfaces B Biointerfaces. 2016; **139**:156-163

[112] Min B-M, Lee G, Kim SH, Nam YS, Lee TS, Park WH. Biomaterials. 2004;**25**: 1289-1297

[113] Kasoju N, Bora U. Silk fibroin in tissue engineering. Advanced Healthcare Materials. 2012;**1**:393-412

[114] Sun K, Li H, Li R, Nian Z, Li D, Xu C. Silk fibroin/collagen and silk fibroin/ chitosan blended three-dimensional scaffolds for tissue engineering. European Journal of Orthopaedic Surgery and Traumatology. 2015;**25**:243-249

[115] Fan Z, Shen Y, Zhang F, Zuo B, Lu Q, Wu P, et al. Control of olfactory ensheathing cell behaviors by electrospun silk fibroin fibers. Cell Transplantation. 2013;**22**:39-50

[116] Di Felice V, Serradifalco C, Rizzuto L, De Luca A, Rappa F, Barone R, et al. Silk fibroin scaffolds enhance cell commitment of adult rat cardiac progenitor cells. Journal of Tissue Engineering and Regenerative Medicine. 2015;**9**:E51-E64

[117] Kogan G, Šoltés L, Stern R, Gemeiner P. Hyaluronic acid: A natural biopolymer with a broad range of biomedical and industrial applications. Biotechnology Letters. 2007;**29**:17-25

[118] Sudha PN, Rose MH. Beneficial effects of hyaluronic acid. Advances in Food and Nutrition Research. 2014;**72**: 137-176

[119] Trombino S, Servidio C, Curcio F, Cassano R. Strategies for hyaluronic

acid-based hydrogel design in drug delivery. Pharmaceutics. 2019;**11**:407

[120] Dai Prè E, Conti G, Sbarbati A. Hyaluronic acid (HA) scaffolds and multipotent stromal cells (MSCs) in regenerative medicine. Stem Cell Reviews and Reports. 2016;**12**: 664-681

[121] Khan F, Ahmad SR. Polysaccharides and their derivatives for versatile tissue engineering application. Macromolecular Bioscience. 2013;**13**:395-421

[122] Gong Y, Han GT, Zhang YM, Zhang JF, Jiang W, Tao XW, et al. Preparation of alginate membrane for tissue engineering. Journal of Polymer Engineering. 2016;**36**:363-370

[123] Sabra W, Zeng A-P, Deckwer W-D. Bacterial alginate: physiology, product quality and process aspects. Applied Microbiology and Biotechnology. 2001; **56**:315-325

[124] Bhattarai N, Li Z, Edmondson D, Zhang M. Alginate-based nanofibrous scaffolds: Structural, mechanical, and biological properties. Advanced Materials. 2006;**18**:1463-1467

[125] Oksman K, Skrifvars M, Selin J-F. Natural fibres as reinforcement in polylactic acid (PLA) composites. Composites Science and Technology. 2003;**63**:1317-1324

[126] Pang X, Zhuang X, Tang Z, Chen X. Polylactic acid (PLA): Research, development and industrialization. 2010. pp. 1125-1136

[127] Jem KJ, van der Pol JF, de Vos S. Microbial lactic acid, its polymer poly (lactic acid), and their industrial applications. In: Plastics from Bacteria. Berlin/Heidelberg, Germany: Springer; 2010. pp. 323-346

[128] Athanasiou KA, Niederauer GG, Agrawal CM. Sterilization, toxicity, biocompatibility and clinical applications of polylactic acid/polyglycolic acid copolymers. Biomaterials. 1996;**17**: 93-102

[129] Ruan R, Vasile C. Polylactic acid (PLA) synthesis and modifications: A review related papers. DOI: 10.1007/ s11458-009-0092-x

[130] Tiwari AP, Hwang TI, Oh J-M, Maharjan B, Chun S, Kim BS, et al. pH/ NIR-responsive polypyrrole-functionalized fibrous localized drug-delivery platform for synergistic cancer therapy. ACS Applied Materials & Interfaces. 2018;**10**:20256-20270

[131] Sun H, Mei L, Song C, Cui X, Wang P. The in vivo degradation, absorption and excretion of PCL-based implant. Biomaterials. 2006;**27**:1735-1740

[132] Sung H-J, Meredith C, Johnson C, Galis ZS. The effect of scaffold degradation rate on three-dimensional cell growth and angiogenesis. Biomaterials. 2004;**25**:5735-5742

[133] Guo B, Ma PX. Synthetic biodegradable functional polymers for tissue engineering: A brief review. Science China. Chemistry. 2014;**57**: 490-500

[134] Shalumon KT, Anulekha KH, Chennazhi KP, Tamura H, Nair SV, Jayakumar R. Fabrication of chitosan/ poly(caprolactone) nanofibrous scaffold for bone and skin tissue engineering. International Journal of Biological Macromolecules. 2011;**48**:571-576

[135] Gunatillake PA, Adhikari R, Gadegaard N. Biodegradable synthetic polymers for tissue engineering. European Cells & Materials. 2003;**5**:1-16

[136] Chen S-H, Tsao C-T, Chang C-H, Lai Y-T, Wu M-F, Chuang C-N, et al. Assessment of reinforced poly(ethylene glycol) chitosan hydrogels as dressings in a mouse skin wound defect model. Materials Science and Engineering: C. 2013;**33**:2584-2594

[137] Zhu KJ, Xiangzhou L, Shilin Y. Preparation, characterization, and properties of polylactide (PLA)–poly (ethylene glycol)(PEG) copolymers: A potential drug carrier. The Journal of Applied Polymer Science. 1990;**39**:1-9

[138] Chai Y, Lin D, Ma Y, Yuan Y, Liu C. RhBMP-2 loaded MBG/PEGylated poly (glycerol sebacate) composite scaffolds for rapid bone regeneration. Journal of Materials Chemistry B. 2017;**5**:4633-4647

[139] Pegg CE, Jones GH, Athauda TJ, Ozer RR, Chalker JM. Facile preparation of ammonium alginate-derived nanofibers carrying diverse therapeutic cargo. Chemical Communications. 2014; **50**:156-158

[140] Zhang X, Reagan MR, Kaplan DL. Electrospun silk biomaterial scaffolds for regenerative medicine. Advanced Drug Delivery Reviews. 2009;**61**:988-1006

[141] Bondar B, Fuchs S, Motta A, Migliaresi C, Kirkpatrick CJ. Functionality of endothelial cells on silk fibroin nets: comparative study of micro- and nanometric fibre size. Biomaterials. 2008;**29**:561-572

[142] Soffer L, Wang X, Zhang X, Kluge J, Dorfmann L, Kaplan DL, et al. Silk-based electrospun tubular scaffolds for tissue-engineered vascular grafts. Journal of Biomaterials Science, Polymer Edition. 2008;**19**:653-664

[143] Pezeshki M, Mojgan M, Sarah Z. Tailoring the gelatin/chitosan electrospun scaffold for application in skin tissue engineering: An in vitro study. Progress in Biomaterials. 2018;**7**: 207-218

[144] Xian J, Ling L, Liau L, Saim A, Yang Y. Electrospun collagen nanofibers and their applications in skin tissue engineering. Tissue Engineering and Regenerative Medicine. 2017. DOI: 10.1007/s13770-017-0075-9

[145] Jin G, Prabhakaran MP, Ramakrishna S. Stem cell differentiation to epidermal lineages on electrospun nanofibrous substrates for skin tissue engineering. Acta Biomaterialia. 2011;**7**: 3113-3122

[146] Seok C, Sook KÆ, Park Y, Jeong ÆH, Jung KÆH. Development of 3-D nanofibrous fibroin scaffold with high porosity by electrospinning: implications for bone regeneration. 2008. pp. 405-410

[147] Yin Z, Chen X, Song H, Hu J, Tang Q, Zhu T, et al. Electrospun scaffolds for multiple tissues regeneration in vivo through topography dependent induction of lineage specific differentiation. Biomaterials. 2015;**44**: 173-185. DOI: 10.1016/j. biomaterials.2014.12.027

[148] Delgado-rangel LH, Hernández-vargas J, Becerra-gonzález M. development of collagen/poly (vinyl alcohol)/chondroitin sulfate and collagen/poly(vinyl alcohol)/ha electrospun scaffolds for tissue engineering. 2019;**20**:2470-2484

[149] Flaig F, Ragot H, Simon A, Revet G, Kitsara M, Kitasato L, et al. Design of functional electrospun scaffolds based on poly (glycerol sebacate) elastomer and poly (lactic acid) for cardiac tissue engineering. ACS Biomaterials Science & Engineering. 2020;**6**:2388-2400

[150] Vogt L, Rivera LR, Liverani L, Piegat A, El Fray M, Boccaccini AR. Poly (ε-caprolactone)/poly(glycerol sebacate) electrospun scaffolds for cardiac tissue engineering using benign solvents. Materials Science and Engineering: C. 2019;**103**:109712

[151] Wright LD, Andric T, Freeman JW. Utilizing NaCl to increase the porosity of electrospun materials. Materials Science and Engineering: C. 2011;**31**:30-36

[152] Blakeney BA, Tambralli A, Anderson JM, Andukuri A, Lim D-J, Dean DR, et al. Cell infiltration and growth in a low density, uncompressed three-dimensional electrospun nanofibrous scaffold. Biomaterials. 2011; **32**:1583-1590

[153] Rnjak-Kovacina J, Weiss AS. Increasing the pore size of electrospun scaffolds. Tissue Engineering. Part B, Reviews. 2011;**17**:365-372

[154] Costa PF, Vaquette C, Zhang Q, Reis RL, Ivanovski S, Hutmacher DW. Advanced tissue engineering scaffold design for regeneration of the complex hierarchical periodontal structure. Journal of Clinical Periodontology. 2014; **41**:283-294

[155] Khorshidi S, Solouk A, Mirzadeh H, Mazinani S, Lagaron JM, Sharifi S, et al. A review of key challenges of electrospun scaffolds for tissue-engineering applications. Journal of Tissue Engineering and Regenerative Medicine. 2016;**10**:715-738

Chapter 3

Production of Nanofibers from Plant Extracts by Electrospinning Method

Nilşen Sünter Eroğlu

Abstract

The fact that different plants grow in each climate type, that each plant has different and many benefits, and that it can obtain bio-structured, sustainable, economic, and ecological products has increased the work of researchers in this field. The long-term toxicity and harmful side effects of herbal extracts are generally less compared to synthetic drugs. Studies on the production of nanofibrous membrane structures from plant extracts are relatively limited and are an emerging field. Herbal extracts have a positive effect in electrospinning applications with their biodiversity, ability to maintain biological functionality, and wound healing effects against pathogenic microorganisms. With the creation of nanofiber structures of plants obtained from natural sources, applications in fields such as wound healing, tissue engineering, drug release are increasing day by day.

Keywords: nanofiber, polymer, electrospinning, herbal extract, electrospun

1. Introduction

Electrospinning is the most preferred method because of its low cost compared to nanofiber production methods, production of long and continuous nanofibers, controllable nanofiber diameter, and industrial processing potential. When all these properties are evaluated, it would be appropriate to produce a nanofiber for wound healing by electrospinning. On the other hand, in recent years, interest in polymer materials obtained by the electrospinning method has increased significantly. Materials such as polymers and nanofiber composites can be produced directly by electrospinning. The post-processing of electrospun fibers forms other materials, such as ceramics and carbon nanotubes [1]. Polymer nanofibers obtained by the electrospinning method have a high surface area-volume ratio, flexible in surface functions, have superior mechanical performance, and are versatile in design [2].

Because of all these advantages, the most common and simple method used for tissue framework production is electrospinning. The principle of operation is based on filling the syringe with the polymer solution or melting in the high potential area and spraying it from the tip of the syringe to the collector by applying a voltage to an electrode connected to the tip of the syringe (**Figure 1**). Here, since the solution sprayed

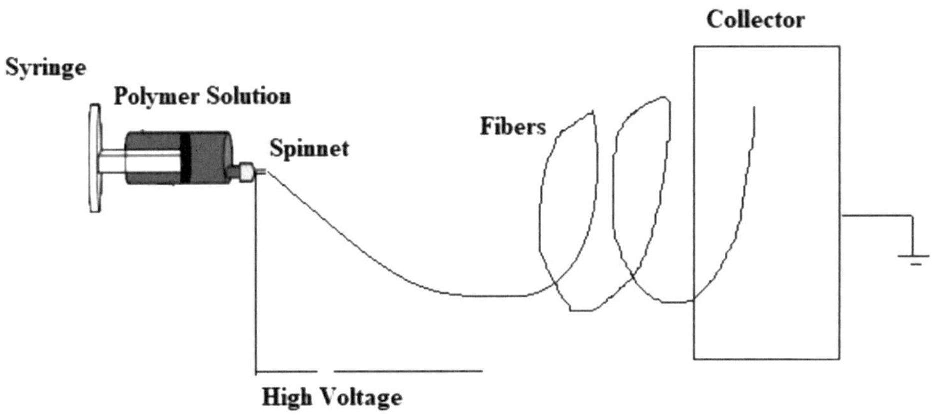

Figure 1.
Schematic representation of the electrospinning process.

from the syringe is subjected to an electrical field, it elongates at the tip of the needle, and a conical appearance called a Taylor cone is obtained. A typical electrospinning process must be between a high voltage source with positive or negative polarity and a grounded surface so that the fibers can clump together. Spraying the solution in the syringe starts when the potential difference applied from the voltage source reaches the threshold value and equalizes to the electrostatic forces, and is completed by spraying it on the grounded surface. Since the fibers collected on the surface are sprayed with a high amount of pulling, they should be in a fine and regular structure [3–5].

The surface tension of the liquid (γ), and the gravitational force (Fg) affect the droplet when the solution, which is the first step of the electrospinning process, comes out of the syringe by forming a droplet. The capillary of internal radius (R), density of the liquid (ρ), and gravitational constant (g) values of the pipe through which the polymer flows are effective in the formation of the radius of the droplet (r0).

$$(r_o) = (3R\gamma / 2\rho g)^{1/3} \quad (1)$$

When a sufficiently high voltage is applied, the electric force FE, the gravitational force Fg encounter surface forces ($F\gamma = FE + Fg$), and the radius of the droplet decreases from (r0) to r ($r < r0$) [6].

After droplet formation, the polymer solution overcomes the surface forces under the influence of Coloumb repulsive forces, forming a Taylor cone with an apex angle of 49.3°. Initially straight, the jet segment may become unstable over time and may show twisting and undulating movements as it passes toward the collector. The jet in this region exhibits components of predominantly non-axial electrostatic repulsion forces. Three types of instability can occur as demonstrated by the polymer jet. These instability forms are listed as classical Rayleigh instability, axisymmetric electric field current, and whipping instability. Whipping instability results in a radial torque from the center of the jet, resulting in a high degree of bending instability. The resulting radial jets push each other and separate from the main jet. The interaction between increasing charge density on the one hand and viscous and surface tension forces resisting elongation on the other determines the complexity of the resulting instability [6, 7].

This chapter focused that the electrospinning process, parameters affecting the process such as solution and ambient. After then, it was explained herbal extracts were used to obtain nanofibers by electrospinning method and their application areas. This chapter will provide an overview of the principles of the electrospinning process with various herbal extracts for potential applications in many fields especially biomedical areas.

2. Parameters affecting electrospinning process

There are three main parameters of the electrospinning process. These are due to the polymer solution, process, and environmental conditions. In this section, the factors affecting each parameter will be discussed in detail. These parameters and their effects in **Table 1** are also shown.

2.1 Solution parameters

To carry out the electrospinning process, the polymer must be in liquid form, in the form of a molten polymer or polymer solution. The physical and chemical properties of the solutions play an active role in the electrospinning process and the resulting fiber morphology. During the electrospinning process, the polymer solution is drawn from the tip of the needle. For this reason, the electrical properties, surface tension, and viscosity of the solution determine the amount of stress in the solution. The evaporation rate also affects the viscosity of the solution as it is stretched. The solubility of the polymer in the solvent determines not only the viscosity of the solution but also the types of polymers that can be mixed with each other [1].

2.1.1 Solution viscosity, molecular weight, and concentration

Viscosity is the most important factor determining the flow rate of the solution. In the electrospinning process, the flow rate increases at low viscosity [8]. However, when the viscosity of the solution is too low, fluidity may occur and polymer particles may form instead of fibers. In solutions with lower viscosity, the polymer chain is generally less synthesized with each other [9], less chain entanglement occurs, and thus jet stability is lost. The fibers are collected into the collector as droplets, which

Parameters	Effect of Fiber Morphology
Solution Viscosity ↑	Fiber Diameter ↑ (within the optimum range)
Surface Tension↓	Fiber Diameter ↑
Solution Conductivity ↑	Fiber Diameter ↓(wide diameter distribution)
Solution Dielectric Constant↑	Fiber Diameter ↓
Voltage ↑	Fiber Diameter ↓ later ↑
Flow Rate ↑	Fiber Diameter ↑ (if the flow rate is too high, a bead appearance occurs.)
Temperature↑	Fiber Diameter ↓ (as the viscosity will decrease)
Distance Between Tip and Collector↑	Fiber Diameter ↓ (if the distance between the tip and the collector is too short, a bead appearance occurs)
Humidity (Moisture) ↑	Fiber Diameter ↑ (with the optimum range)

Table 1.
Electrospinning parameters affecting fiber morphology.

first turn into spindle-like structures and then into beaded nanofibers [3]. As the viscosity increases, the formation of the bead structure decreases, and more regular nanofibers are obtained [9]. Therefore, factors that affect the viscosity of the solution also affect the electrospinning process and the resulting fibers.

The molecular weight of the polymer used in the electrospinning process has a direct effect on properties such as viscosity, surface tension, and conductivity, and this interaction determines the nanofiber formation. Molecular weight is explained as the length of the chains of the polymer from which nanofibers will be obtained [4]. The length of the polymer chain will determine the amount of entanglement of the polymer chains in the solvent [1]. Since the viscosity will be higher in polymer solutions with high molecular weight, the formation of beads decreases [10]. Although the increase in molecular weight provides regular fiber formation, if this increase is high, it causes the formation of microstrip structure [11, 12].

2.1.2 Surface tension

When a very small drop of waterfalls into the air, the droplet usually takes on a spherical shape. The liquid surface property that causes this phenomenon, which occurs when the electrical forces are around zero, is known as surface tension [1]. An excessive increase in surface tension adversely affects the electrospinning process. Some surfactants with low concentrations are used to lower the surface tension. The decrease in the surface tension of the solution ensures the formation of finer and smoother fibers and a problem-free electrospinning process [4]. The concentration change in the solutions used directly affects the surface tension [13].

2.1.3 Solution conductivity

Electrospinning is a method of obtaining nanofibers that repel the charges on the surface by stretching the solution and transfer the electric charge from the electrode to the polymer solution [1, 14]. In the electrospinning system, low electrical conductivity can form beaded fibers as it will create instability and cause the jet to not be able to extend sufficiently, while with high electrical conductivity, the polymer jet can stretch more with the loads it carries and form fibers with a smoother and finer structure [3]. For this reason, it is aimed to increase the electrical conductivity by increasing the concentration. Some additives can also be added to increase conductivity in low-concentration or insufficiently ionic solutions [4]. If the conductivity of the solution increases, the electrospinning jet can carry more charge. For example, the conductivity of the solution can be increased by the addition of ions [1]. By adding salt to an uncharged solution, although electrical neutrality is maintained, salt molecules can dissociate into independently acting positive and negative ions, thereby increasing the electrical conductivity of a solution [15, 16]. These ions can also be obtained by dissolving most drugs or proteins in water. As a result, when a small amount of salt or polyelectrolyte is added to the solution, the increased loads carried by the solution will increase the stretching of the solution and the formation of beaded fibers will be prevented [1].

2.1.4 Solution dielectric constant

The insulation constant or dielectric constant is defined as a coefficient that measures the ability of a material to store charge on it [17]. As the dielectric constant of the solutions increases, the charge distribution across the surface of the bubble formed at the needle tip will be more uniform, as there will be more net charge density. Therefore, the ordered structure of the obtained nanofiber is also increasing [3, 4]. It is thought that as the dielectric constant increases, obtaining finer and smoother fibers is due to the application of more tension force to the fluid jet [18].

2.2 Processing variables

Another important parameter affecting the electrospinning process is various external factors applied to the electrospinning jet. These factors are voltage, flow rate, temperature, collector effect, nozzle diameter, and the distance between the tip and the collector. Although these parameters are less important than solution parameters, they have a certain effect on fiber morphology.

2.2.1 Voltage

Voltage is a parameter that induces charges in the solution, overcomes electrostatic forces, and initiates the electrospinning process [1]. As the amount of applied voltage increases, the diameters of the obtained nanofibers will decrease [4]. There are three main reasons for this. The first reason is because of a higher voltage will lead to greater stretching of the solution due to the larger columbic forces in the jet and the stronger electric field. This will reduce the fiber diameter. The second factor is that by using a lower viscosity solution, at a higher voltage, the formation of secondary jets during electrospinning is achieved. Thus, the fiber diameters can become narrower. Another factor that can affect the fiber diameter is the flight time of the electrospinning jet. A longer flight time will allow more time for the fibers to stretch and elongate before being placed on the collecting plate. Therefore, at a lower voltage, the diminished acceleration of the jet and the weaker electric field can increase the flight time of the electrospinning jet, facilitating the formation of finer fibers [1].

In many studies [19–21], it was observed that the formation of beads on the surface formed with the increase of voltage increased. The increase in bead density due to tension is explained because of increased instability of the jet as it is drawn into the syringe needle in the Taylor cone [1]. Here, bead formation occurs with the excessive acceleration of voltage increase, jet movement, and evaporation [4]. It is also suggested that increasing voltage will increase bead density and at even higher voltage, beads will form fibers of thicker diameter [1].

Despite these studies that the voltage increase creates a bead surface, it has been observed that the production of nanofibers at very low voltage also creates a beaded surface [22]. In this sense, the important thing is to work at a voltage where the flow balance will be stable. With the increase in voltage, the jets coming out of the cone tip reach the collector in an orderly manner, increasing their speed in the electrical field. Here, excessive speed increase or decrease is a factor that will lead to the formation of a beaded surface. In other words, the applied voltage must have an upper and lower limit.

2.2.2 Feeding rate

The feeding rate determines the amount of feed in the electrospinning system. A certain feeding rate is needed to maintain the Taylor cone in the system. When the feeding rate increases, there will be an increase in the fiber diameter or the size of the beads formed in the fibers, as there will be more solution volume at the nozzle tip [1].

At low feeding rate, nanofiber production will not be possible because there will not be sufficient feed for the Taylor cone.

As the applied voltage changes, the resulting Taylor cone will also change. At low applied voltages, a hanging drop forms at the tip of the array. The Taylor cone is then formed at the tip of the array. However, as the applied voltage increases (moving from left to right), the volume of the hanging drop decreases until a Taylor cone is formed at the tip of the array. Increasing the applied voltage results in the ejection of the spray through the syringe, which is associated with an increase in bead formation [5].

2.2.3 Temperature

The temperature parameter consists of three environmental variables: melt temperature, solution temperature, and ambient temperature. As the melt temperature increases, less tension is required due to the decrease in viscosity and fiber diameters decrease [7]. Similar to melt temperature, the temperature of a solution has the effect of both increasing the evaporation rate and reducing the viscosity of the polymer solution. This is because the solution has a lower viscosity and greater solubility of the polymer in the solvent, allowing the solution to be stretched more evenly. With a lower viscosity, Columbic forces can exert a greater tensile force on the solution, thus resulting in smaller diameter fibers [1].

2.2.4 Effect of the collector

There must be an electric field between the source and the collector (collector) for the electrospinning process to start. Therefore, in most electrospinning systems, the collector plate is made of a conductive material such as aluminum foil, which is electrically grounded such that there is a constant potential difference between the source and the collector. If a non-conductive material is used as a collector, charges from the electrospinning jet will quickly build upon the collector, resulting in less fiber deposition. Fibers collected on non-conductive material generally have a lower packing density than those collected on a conductive surface. This is due to the repulsive forces of the loads that build upon the collector as more fibers accumulate. For a conductive collector, the loads on the fibers are distributed so that more fibers are drawn into the collector. As a result, the fibers can be wrapped closely together [1].

The most commonly used collector types in the electrospinning method are generally flat plates, grids and frames. Apart from these, rotating cylinder, rotating disc, rotating cones, parallel rings, liquid bath and wrapper, pyramid-shaped platform, conveyor belt, two parallel frames, rotor, and thin conductive rod are listed as [7].

2.2.5 Nozzle (needle) diameter

The nozzle diameter has a certain effect on the electrospinning process. As the nozzle diameter gets smaller, it provides clogging of the diameter and reduces the amount of beads on the nanofibers. The reduction in occlusion is due to less exposure of the solution to the atmosphere during electrospinning. The decrease in the inner diameter of the hole causes a decrease in the diameter of the nanofibers. As the size of the droplet at the tip of the hole decreases, the surface tension of the droplet increases. It reduces jet acceleration when the same amount of voltage is applied, allowing more time for the solution to stretch and stretch before the collector. The nanofibers formed in this way are finer [1].

The nozzle could be blockage when electrospinning with electrospinning chloroform solutions of PLA. When more than one nozzle is formed, the solvent density may increase, but this will increase the difficulty of solvent removal and nozzle cleaning and compose the deposition of nonwoven fiber in thicknesses >10 mm [23].

2.2.6 Distance between tip and collector

The distance between the needle tip and the collector provides the necessary time for the solvent in the polymer jet sprayed from the nozzle tip to evaporate [4]. Changing the distance between the tip and the collector has a direct effect on both the flight time and the electric field strength. As the distance between the tip and the collector decreases, the jet will have a shorter distance to travel before reaching the collector plate. In addition, the electric field strength will increase at the same time, which will increase the acceleration of the jet going to the collector. As a result, there may not be enough time for solvents to evaporate when they hit the collector. When the distance is too low, excess solvent causes the fibers to coalesce where they come into contact [1].

2.3 Environmental conditions

Environmental conditions in the electrospinning process are the factors affecting the electrospinning process. In this sense, humidity, atmospheric type, and pressure cause physical and morphological changes in the formed fibers.

2.3.1 Humidity (moisture)

The increase in humidity in the environment adversely affects the electrospinning process. High humidity causes circular pores to form on the nanofiber surfaces obtained. The pore depth increases with increasing humidity. However, the depth, diameter, and number of pores remain constant above a certain humidity [1]. It is not possible to carry out the electrospinning process at very high humidity values [3]. As the humidity level decreases, volatile solvents evaporate quickly, causing drying and making the electrospinning process difficult [9]. For this reason, keeping the humidity level at an optimum level is an important factor.

2.3.2 Type of atmosphere

The type of atmosphere in which the electrospinning process takes place is very important for the smooth running of the process. Different gases have different behavior under the high electrostatic field. For example, helium decomposes under a high electrostatic field and therefore electrospinning is not possible [1]. For another example, with excessively volatile solvents the Taylor cone could dry out. To prevent evaporation in the cone, it is feasible to introduce a local stream of solvent-saturated gas around the cone [23]. The decrease in pressure in the environment adversely affects the electrospinning process [1].

2.3.3 Pressure

Pressure changes in the electrospinning process make it difficult to ensure the stability of the drafting process. The reduction in pressure surrounding the electrospinning jet adversely affects the electrospinning process. When the ambient pressure drops below atmospheric pressure, the polymer solution in the syringe will have a greater tendency to flow through the needle, resulting in an unstable spray start. As the pressure decreases, rapid solution foaming occurs at the needle tip. At very low pressure, electrospinning is not possible due to the direct discharge of electric charges [1].

3. Herbal extracts used to obtain nanofibers by electrospinning method

Reasons such as health problems, population density, environmental pollution, and increased consumption have encouraged people to seek natural solutions. The use of herbal products in the field of health for their healing properties is increasing day by day. In recent years, plants derived from natural substances such as flavonoids, terpenoids, steroids have received considerable attention due to their different pharmacological properties, including antibacterial, antioxidant, and anticancer activity.

The olive leaf plant, which draws attention with its biocompatible, biodegradable, antioxidant, and antimicrobial properties, has been used by many researchers [24–27] in the electrospinning process for use in the biomedical field. Similarly, because of its biocompatible, biodegradable and antimicrobial properties, and rosemary plant [28, 29] has been used as a bioactive packaging material and to obtain nanofibers by electrospinning for use in the biomedical field. Many plant extracts such as aloe vera [30, 31], thyme [10], grape seed [32], chamomile [33], green tea [34], grewia mollis [35], gotu kola [36], calendula [37], mangosteen [38], lavender [39] are mixed with different polymers and used in the production of nanofibers for use in the medical field.

4. Application areas of nanofibers obtained from plant extracts by electrospinning method

The fact that different plants grow in every geography, each plant has different and many effects and the ability to obtain biocompatible, sustainable, organic, and environmentally friendly products have encouraged researchers to work in this field. The use of plants obtained from natural sources as active agents is increasing day by day in areas such as wound healing, tissue engineering, and drug release.

4.1 Wound healing

The skin forms the largest part of body weight and is very vulnerable to external forces and effects such as tissue traumas and injuries. Today, wound dressings play a vital role in the healing of such wounds, and wound healing depends on several factors such as selection of wound dressing, physiological state of the wound, and degree of damage. An ideal wound dressing should facilitate wound healing, remove exudates from the wound bed, be non-toxic and allergenic, and act as a barrier against microbes [4, 30]. Conventional wound dressings are generally used to close the wound and absorb the excess discharge. Although in previous studies, it was stated that the dressing should keep the wound dry, it is known that a warm and moist environment on the wound increases the healing of the wound [40]. However, it is a fact that excessive moisture causes wetting and softening of the scar tissues and prolongs the wound healing process [4]. Keeping the humidity level at an optimum level is very important for wound treatment. In addition to the ideal moisture level of modern wound dressings, effective oxygen circulation, air permeability, and low bacterial contamination are the essential qualities sought [40].

Modern wound dressings are composed of water-absorbent granular hydrocolloids, alginate containing mannuronic and guluronic acids, and hydrogel, in which water-absorbing polymers are structured into a three-dimensional network [40]. In recent years, with the rapid development of tissue engineering, nanofiber-based ECM (extracellular matrix) scaffold structures have become widespread [4]. ECM is a collagenous substance commonly found in skin, tendons, cartilage, and bone [11]. Compared to other wound dressings, nanofiber wound dressings have advantages such as hemostasis, high porosity, good fluid absorption capacity, small pore sizes, and large surface area [4]. Hyaluronic acid, collagen, chitosan-based nanofibers are generally used in new generation nanofiber-containing bioactive wound dressings due to their biocompatible, biodegradable, and antibacterial properties [40]. Thus, it ensures the healing of the wound by releasing the active substance in the nanofiber structure onto the wound in a controlled manner.

In recent studies [24, 29, 41–43], herbal extracts seem to be helpful in fighting infection and accelerating the wound healing process. The use of herbal extracts as wound dressings can nourish the wound site with healing properties such as antimicrobial, anti-inflammatory, analgesic, and tissue regeneration [30]. The long-term toxicity and harmful side effects of herbal extracts are generally insignificant compared to synthetic drugs. The main disadvantage of herbal medicines is that they need to be used in higher dosages than synthetic medicines. Large amounts of herbal medicines extracted from plants reduce their solubility in water or other chemical solvents. Therefore, dissolution of plant extracts almost never occurs in polymer-carriers such as capsules, nanofiber mats, and casting films containing herbal medicines. This may cause adverse effects in applications such as drug release behavior. Despite these problems, herbal drugs promise great success compared to chemical drugs due to their superior performance in wound treatments [10]. The important point here is to extract the herbal extracts in a suitable solvent, to obtain a biocompatible polymer and a nanofibrous structure that preserves its existing effects such as anti-inflammatory and antibacterial and supports the repair of opened wounds.

4.2 Tissue engineering

Tissue engineering is a field that aims to heal damaged or diseased tissues/organs, to maintain, regenerate and develop the functions of normal tissues/organs, and to form tissue scaffolds with repair capability for this purpose. Electrospinning is an application with high potential in many tissue engineering fields such as vasculature, bone, neural, and tendon/ligament. With the electrospinning process, the ability to form aligned scaffolds for anisotropic mechanical and biological properties in the field of vascular grafts, as well as the ability to inhibit smooth muscle cell migration, is provided. In addition, possibilities have been presented to improve vascular grafts with tissue scaffolds that can be obtained by tissue engineering [5, 44].

Nanofibers in tissue engineering must have such as biocompatible, biodegradable (with an acceptable shelf life), tissue-appropriate degradation rate, tissue-appropriate mechanical (strength, stiffness, and modulus) and structural (pore sizes, shape, and structure) properties, and sterilizability [45]. Tissues consist of multiple cell types and works in conjunction with the cell-surrounding extracellular matrix (ECM), which is the tissue scaffold, concealed by regular, micro-sized cells. The ECM is responsible for providing the cells with the needed mechanical support and protecting the cells. The materials used in tissue engineering applications should allow a certain interaction with the cell, the cell's attachment, proliferation, change, ECM production, and proper progression of this process should be ensured. It should form a supporting function in the formation of new tissues [3, 44].

Approximately, 25% of current prescription drugs are derived from trees, medicinal plants, shrubs, and herbs in nature. The use of herbal extracts with nanofibers produced by electrospinning provides a good potential to form scaffolds for skin regeneration [46]. For example, it has been seen that the nanofibrous structure of the chamomile plant supports collagen fiber accumulation and tissue formation in the dermis [33], and the olive leaf plant has a good potential for tissue scaffolding in biomedical applications thanks to its high antioxidant effect [3]. There are studies on tissue scaffolds containing edible, non-toxic, biocompatible, biodegradable plant extracts with many different contents. It is thought that the applications of plant-based tissue scaffolds will increase in future studies.

4.3 Drug delivery systems

Drug delivery systems aim to deliver the drug to the unhealthy region in a controlled and regular manner and to ensure its effectiveness in this region. While drug delivery is generally associated with the delivery of therapeutic agents for the treatment of certain disease states such as cancer, the delivery systems for tissue engineering applications can also apply to the delivery of bioactive agents such as proteins and DNA [5].

In conventional drug delivery systems, successive doses of the drug cause a fluctuating profile of the drug concentration in the blood throughout the treatment period. Therefore, at certain times, concentrations may exceed the recommended maximum (Cmax) concentration with the risk of biotoxicity or fall below the minimum concentration (Cmin), limiting the therapeutic effect. To obtain the highest therapeutic value from the drug, the optimum concentration (C), (Cmin < C < Cmax) in body tissue should be maintained throughout the entire treatment period. Via controlled delivery techniques, the bioavailability of the drug has been designed throughout to be close to this optimum value. In addition, the amount of drug required to be administered is relatively lower in the controlled release mode, minimizing potential side effects [6].

In tissue engineering, the design of the polymer scaffold requires the release of growth factors and other bioactive substances into the growing tissue over a period of time. In nanofiber applications such as wound dressings or artificial leather, the local controlled release of antibiotic substances can aid the healing process. Polymer-based delivery systems can produce controlled drug release by diffusion or chemical bioerosion of the matrix or biodegradation of the linkages connecting the drug to the matrix [6]. These advantages are of great importance in their preference and use.

Polymer-based drug delivery systems; nano or microparticles, hydrogels, micelles, and fibrillar systems. Fibrillated systems form nanofiber-based drug release systems [3]. The release kinetics of the drug is controlled by the morphology of the polymer/drug composite as well as the semi-crystalline structure of the polymer. First, the drug is dissolved at the molecular level in the polymer matrix. The drug is separated as crystalline or amorphous particles in the polymer matrix [6, 47].

The use of herbal-based nanofiber structures in drug delivery systems has increased in recent years. There are different applications such as designing coaxial nanofibers by using olive leaf extract as a bioactive agent [25], producing nanofiber membranes containing aloe vera [48], using nanofibers prepared using the bark of

Tecomella undulate (rohida) plant in in-vitro drug release [49]. It is expected that nanofiber drug delivery systems containing herbal extracts will increase therapeutic efficacy, reduce toxicity and ensure compatibility with patients by delivering drugs to the affected area at a controlled rate for a certain period.

5. Conclusion

Electrospinning is a nanofiber production method that is the most preferred because it is simple, economical, and environmentally friendly, and has many production parameters including solution, process, and environmental conditions. Production of nanofibers by electrospinning process; It is a subject that draws attention with its applications in many fields such as tissue engineering, drug release, filtration, automotive, energy, food industry, cosmetics, agriculture, biosensing. Although polymer contents with synthetic infrastructures are generally preferred in these applications, approaches to using natural agents with few side effects, biocompatible, sustainable, economical, biodegradable, and free from toxic components are increasing [10, 27, 33, 41, 42]. The use of natural components containing active agents in the production of nanofibers is becoming more and more common in the fight against potential health problems that may occur due to the rapidly increasing world population and environmental pollution. Herbal extracts are promising in electrospinning applications with their biodiversity, ability to maintain their biological functionality even after exposure to high electrical voltage, and wound healing effects against pathogenic microorganisms. In addition, it is thought that the use of herbal extracts in different applications in the field of health will become widespread, as they have fewer side effects, and versatile therapeutic properties compared to chemical agents.

Conflict of interest

The authors declare no conflict of interest.

Author details

Nilşen Sünter Eroğlu
Haliç University, İstanbul, Turkey

*Address all correspondence to: nilseneroglu@halic.edu.tr

References

[1] Ramakrishna S, Fujihara K, Teo WE, Lim TC, Ma Z. An Introduction to Electrospinning and Nanofibers. Singapore: World Scientific, National University; 2005

[2] Gazioğlu RD. Keratin esaslı yüzeylerin elektroçekim yöntemiyle elde edilmesi, karakterizasyonu ve gaz sorpsiyon özelliklerinin incelenmesi, [PhD thesis]. Türkiye: Bursa Teknik Üniversitesi, Bursa; 2018

[3] Doğan G. Elektrolif Çekim Yöntemiyle Elde Edilen Biyopolimer Nanoliflerin Doku Mühendisliği ve ilaç Salımı Uygulamalarında Kullanım Olanaklarının Araştırılması. [PhD thesis]. İzmir, Türkiye: Ege Üniversitesi; 2012

[4] Erdem R. Nanolif Bazlı Yara Örtüsü Yüzeyi Geliştirilmesi. İstanbul, Türkiye: Marmara Üniversitesi Fen Bilimleri Enstitüsü [PhD thesis]; 2013

[5] Sill TJ, von Recum HA. Electrospinning: Applications in drug delivery and tissue engineering. Biomaterials. 2008;**29**(13):1989-2006. DOI: 10.1016/j.biomaterials.2008.01.011

[6] Andrady AL. Science and Technology of Polymer Nanofibers'. Vol. 135. New Jersey: John Wiley & Sons, Inc.; 2008

[7] Kozanoğlu G. Elektrospining Yöntemiyle Nanolif Üretim Teknolojisi, [Msc thesis]. İstanbul, Türkiye: İstanbul Teknik Üniversitesi, Fen Bilimleri Enstitüsü; 2006

[8] Call CC. The Study of Electrospun Nanofibers and the Application of Electrospinning In Engıneering Education, [MSc Thesis]. United States of America: Graduate Studies of Texas A&M University; 2008

[9] Huang Z-M, Zhang Y-Z, Kotaki M, Ramakrishna S. A review on polymer nanofibers by electrospinning and their applications in nanocomposites. Composites Science and Technology. 2003;**63**(15):2223-2253. DOI: 10.1016/S0266-3538(03)00178-7

[10] Karami Z, Rezaeian I, Zahedi P, Abdollahi M. Preparation and performance evaluations of electrospun poly(e-caprolactone), poly(lactic acid), and their hybrid (50/50)Nanofibrous Mats containing thymol as an herbal drug forEffective wound healing. Journal of Applied Polymer Science. 2013;**129**:756-766. DOI: 10.1002/app.38683

[11] Tort S. Elektro-Çekim Yöntemiyle Hazırlanan Nanolif Yara Örtülerinin İn Vitro/İn Vivo Değerlendirilmesi. Ankara, Türkiye: Gazi Üniversitesi, Farmasötik Teknoloji Anabilim Dalı, [PhD thesis]; 2016

[12] Koski A, Yim K, Shivkumar S. Effect of molecular weight on fibrous PVA produced by electrospinning. Materials Letters. 2004;**8**(3-4):493-497. DOI: 10.1016/S0167-577X(03)00532-9

[13] Yang Q, Li Z, Hong Y, Zhao Y, Qiu S, Wang C, et al. Influence of solvents on the formation of ultrathin uniform poly(vinyl pyrrolidone) nanofibers with electrospinning. Journal of Polymer Science Part B: Polymer Physics. 2004;**42**(20):3721-3726. DOI: 10.1002/polb.20222

[14] Baumgarten PK. Electrostatic spinning of acrylic microfibers. Journal of Colloid and Interface Science. 1971;**36**(1):71-79. DOI: 10.1016/0021-9797(71)90241-4

[15] Reneker DH, Yarin AL. Electrospinning jets and polymer nanofibers. Polymer.

2008;**49**(10):2387-2425. DOI: 10.1016/j.polymer.2008.02.002

[16] Chronakis IS, Grapenson S, Jakob A. Conductive polypyrrole nanofibers via electrospinning: Electrical and morphological properties. Polymer. 2006;**47**(5):1597-1603. DOI: 10.1016/j.polymer.2006.01.032

[17] Wikipedia. 2020.Available from: https://tr.wikipedia.org/wiki/Yal%C4%B1tkanl%C4%B1k_sabiti [Accessed: September 16, 2020]

[18] Son WK, Youk JH, Lee TS, Park WH. The effects of solution properties and polyelectrolyte on electrospinning of ultrafine poly(ethylene oxide) fibers. Polymer. 2004;**45**(9):2959-2966. DOI: 10.1016/j.polymer.2004.03.006

[19] Zhong XH, Kim KS, Fang DF, Ran SF, Hsiao BS, Chu B. Structure and process relationship of electrospun bioabsorbable nanofiber membranes. Polymer. 2002;**43**:4403-4412. DOI: 10.1016/S0032-3861(02)00275-6

[20] Deitzel JM, Kleinmeyer J, Harris D, Beck Tan NC. The effect of processing variables on the morphology of electrospun nanofibers and textiles. Polymer. 2001;**42**:261-272. DOI: 10.1016/S0032-3861(00)00250-0

[21] Demir MM, Yilgor I, Yilgor E, Erman B. Electrospinning of polyurethane fibers. Polymer. 2002;**43**:3303-3309. DOI: 10.1016/S0032-3861(02)00136-2

[22] Tong H-W, Wang M. Electrospinning of poly(Hydroxybutyrate-co-hydroxyvalerate) fibrous scaffolds for tissue engineering applications: Effects of electrospinning parameters and solution properties. Journal of Macromolecular Science, Part B. 2011;**50**(8):1535-1558. DOI: 10.1080/00222348.2010.541008

[23] Stanger J, Tucker N, Staiger M. Electrospinning, Volume 16, Number 10. Shropshire, United Kingdom: Rapra Technology; 2005

[24] Peršin Z, Ravber M, Kleinschek KS, Knez Z, Kerget MS, Kurečič M. Bio-nanofibrous mats as potential delivering systems of natural substances. Textile Research Journal. 2017;**87**(4):444-459. DOI: 10.1177%2F0040517516631323

[25] Dogan G, Başal G, Bayraktar O, Özyıldız F, Uzel A, Erdogan İ. Bioactive sheath/Core nanofibers containing olive leaf extract. Mıcroscopy Research and Technıque. 2015;**79**:38-49. DOI: 10.1002/jemt.22603

[26] Basal G, Tetik G.D, Kurkcu G, Bayraktar O, Gurhan I. D, Atabey A, Olive leaf extract loaded Silk Fibroın/Hyaluronic acid nanofiber webs for wound dressing applications, Digest Journal of Nanomaterials and Biostructures. 2016;**11**(4):1113-1123

[27] Sünter EN, Canoğlu S, Yüksek M. Characterization, mechanical, and antibacterial properties of nanofibers derived from olive leaf, fumitory, and terebinth extracts. Turkish Journal of Chemistry. 2020;**44**:1043-1057. DOI: 10.3906/kim-2003-45

[28] Bhullar KS, Kaya B, Byung-Guk JM. Development of bioactive packaging structure using melt electrospinning. Journal of Polymers and the Environment. 2015;**23**:416-423. DOI: 10.1007/s10924-015-0713-z

[29] Bhullar SK, Ran D, Kaya Ozsel B, Yadav R, Kaur G, Chintamaneni M, et al. Comparative study of the antibacterial activity of rosemary extract blended with polymeric biomaterials. Journal of Bionanoscience. 2016;**10**:326-330. DOI: 10.1166/jbns.2016.1385

[30] Agnes MS, Giri VRD. Electrospun herbal nanofibrous wound dressings for skin tissue engineering. The Journal of The Textile Institute. 2014;**106**(8):886-895. DOI: 10.1080/00405000.2014.951247

[31] Ahmad MR, Yahya MF, editors. Characteristics of electrospun PVA-Aloe vera Nanofibres produced via electrospinning. In: Proceedings of the International Colloquium in Textile Engineering, Fashion, Apparel and Design 2014 (ICTEFAD 2014). 2014. DOI: 10.1007/978-981-287-011-7_2

[32] Lin S, Chen M, Jiang H, Fan L, Sun B, Yu F, et al. Green electrospun grape seed extract-loaded silk fibroin nanofibrous mats with excellent cytocompatibility and antioxidant effect. Colloids and Surfaces B: Biointerfaces. 2015;**139**:156-163. DOI: 10.1016/j.colsurfb.2015.12.001

[33] Motealleh B, Zahedi P, Rezaeian I, Moghimi M, Abdolghaffari AH, Zarandi MA. Morphology, drug release, antibacterial, cell proliferation, and histology studies of chamomile-loaded wound dressing mats based on electrospun nanofibrous poly (e-caprolactone)/polystyrene blends. Journal of Biomedical Materials Research B: Applied Biomaterials. 2014;**102B**: 977-987. DOI: 10.1002/jbm.b.33078

[34] Sadri M, Arab-Sorkhi S, Vatani H, Bagheri-Pebdeni A. New wound dressing polymeric nanofiber containing green tea extract prepared by electrospinning method. Fibers and Polymers. 2015;**16**(8):1742-1750. DOI: 10.1007/s12221-015-5297-7

[35] Al-Youssef HM, Amina M, Hassan S, Amna T, Jeong JW, Nam KT, et al. Herbal drug loaded poly(D,L-lactide-co-glycolide) ultrafine Fibers: Interaction with pathogenic Bacteria. Macromolecular Research. 2013;**21**(6):589-598. DOI: 10,1007/s13233-013-1062-1

[36] Raharjo AB, Putra RDA, IndayaningsihN, SrifianaY, HardiansyahA, Irmawati Y, et al. Preparation of polyvinyl alcohol/asiaticoside/chitosan membrane nano-composite using electrospinning technique for wound dressing. AIP Conference Proceedings. 2020;**2256**:030023. DOI: 10.1063/5.0014545

[37] Torres VEA, do Vale Baracho NC, de Brito J, de Queiroz AAA. Hyperbranched polyglycerol electrospun nanofibers for wound dressing applications. Acta Biomaterialia. 2010;**6**(3):1069-1078. DOI: 10.1016/j.actbio.2009.09.018

[38] Suwantong O, Pankongadisak P, Deachathai S, et al. Electrospun poly (L-lactic acid) fiber mats containing crude *Garcinia mangostana* extracts for use as wound dressings. Polymer Bulletin. 2014;**71**:925-949. DOI: 10.1007/s00289-014-1102-9

[39] Sequeira RS, Miguel SP, Cabral CSD, Moreira AF, Ferreira P, Correia IJ. Development of a poly(vinyl alcohol)/lysine electrospun membrane-based drug delivery system for improved skin regeneration. International Journal of Pharmaceutics. 2019;**570**:118640. DOI: 10.1016/j.ijpharm.2019.118640

[40] Kurtoğlu AH, Karataş A. Yara Tedavisinde Güncel Yaklaşımlar: Modern Yara Örtüleri, Ankara Ecz. Fak. Derg. J. Fac. Pharm. 2009;**38**(3):211-232

[41] Yao C-H, Yeh J-Y, Chen Y-S, Li M-H, Huang C-H. Wound-healing effect of electrospun gelatin nanofibres containing Centella asiatica extract in a rat model. Journal of Tissue Engineering and Regenerative Medicine. 2015;**11**(3): 905-915. DOI: 10.1002/term.1992

[42] Lin S, Chen M, Jiang H, Fan L, Sun B, Yu F, et al. Green electrospun grape seed extract-loaded silk fibroin nanofibrous mats with excellent cytocompatibility and antioxidant effect. Colloids and Surfaces. B, Biointerfaces. 2016;**1**(139):156-163. DOI: 10.1016/j.colsurfb.2015.12.001

[43] Mirzaei E, Sarkar S, Rezayat SM, Faridi-Majidi R. Herbal extract loaded chitosan-based nanofibers as a potential wound dressing. Journal of Advanced Medical Sciences and Applied Technologies (JAMSAT). 2016;**2**(1):141-150. DOI: 10.18869/NRIP.JAMSAT.2.1.141

[44] Can N, Ersoy M. Nanolif Yapılı Polimerik Doku İskeleleri. Tekstil ve Mühendis. 2014;**21**(95):37-50. DOI: 10.7216/130075992014219505

[45] Doğan G, Basal D. Elektrolif Çekim Yöntemine Göre Elde Edilen Biyopolimer Nanoliflerin İlaç Salınım Sistemleri. Yara Örtüsü ve Doku Đskelesi Olarak Kullanımları, Tekstil Teknolojileri Elektronik Dergisi. 2009;**3**(2):58-70

[46] Jin G, Prabhakaran MP, Kai D, Annamalai SK, Arunachalam KD, Ramakrishna S. Tissue engineered plant extracts as nanofibrous wound dressing. Biomaterials. 2013;**34**(3):724-734. DOI: 10.1016/j.biomaterials.2012.10.026

[47] Düzyer Ş. Elektroçekim yöntemiyle üretilen polister nanoliflerinin medikal alanda kullanılabilirliklerinin araştırılması, [PhD thesis]. Bursa, Türkiye: Uludağ Üniversitesi; 2014

[48] Shukry NAA, Sekak KA, Effendi AMR, T.J.B. Characteristics of electrospun PVA-Aloe vera Nanofibres produced via electrospinning. In: Proceedings of the International Colloquium in Textile Engineering, Fashion, Apparel and Design 2014 (ICTEFAD 2014)

[49] Suganya S, Senthil RT, Lakshmi BS, Giridev VR. Herbal drug incorporated antibacterial Nanofibrous mat fabricated by electrospinning: An excellent matrix for wound dressings. Journal of Applied Polymer Science. 2011;**121**(5):2893-2899. DOI: 10.1002/app.33915

Chapter 4

Biomass Electrospinning: Recycling Materials for Green Economy Applications

Farai Dziike, Phylis Makurunje and Refilwe Matshitse

Abstract

The development and advancement of electrospinning (ES) presents a unique material technology of the future achieved by fabricating novel nanofibrous materials with multifunctional physical (three-dimensional [3D] structure, nanoscalable sizes) and chemical characteristics (functional groups). Advancing the possibility of preparing various classes of novel organic and inorganic electrospun fiber composites with unique features such as polymer alloys, nanoparticles (NPs), active agents, and devices. This feature gives provision for internal access of the setup parameters such as polymer precursor material, polymer concentration, solvent, and the method of fiber collection that consequentially improves the intrinsic control of the construction mechanism of the final nanofibrous architecture. In synthetic electrospinning, the nanofibrous material processing allows for internal control of the electrospinning mechanism and foster chemical crosslinking to generate covalent connections between polymeric fibers. Comparing technologies according to materials of the future revealed that electrospinning supports the formation of micro-scale and in some cases nano-scale fibers while the formation of thin films is facilitated by the electrospraying system. Recent innovations point to various biomass waste streams that may be used as an alternative source of polymeric materials for application in electrospinning to produce materials for the future.

Keywords: electrospinning, biomass, nanofibrous, material mimicking, electrospraying, scaffold, nanostructures

1. Introduction

Electrospinning (ES), as a nanotechnology, exhibits strong evolution of materials used across a broad spectrum from bioactive (microorganisms-infused for biomedical applications) to manufacturing (adhesion, proliferation, and differentiation of the mimetic for mechanical, chemical and electrochemical applications) nanofibers [1]. The advent of bioeconomy and innovation technological development presented opportunities for remarkable progress in the expansion of methods and multiple applicability for the electrospun nanofibers. Waste biomass and other recyclable materials are also finding use in ES as an adaptable and sustainable innovative approach for making ultrathin fibers [2]. Valorization of biomass waste materials such as plant biomass, waste plastic, industrial effluent and other waste biomass streams have been processed through various technologies to produce a wide range of higher hierarchical recycled fibrous products. These including biodegradable bio plastic, filtration membranes, nanofibers as macro, micro and nanomaterials. Advancement in innovative ES techniques allows for intrinsic control of the physicochemical factors, including physical (morphology, diameter, orientation); surface (volumetric dispersion, porosity and thickness) and chemical (functional groups) characteristics of the final product [2].

Electrospinning method entails the utilization of voltage to create an electric field, polymer solution of specific concentration and electrospinning pump to introduce the spinneret onto collector plate. The resulting products are electrospun nanofibers characterized by their fibrous morphology, three-dimensional (3D) porous framework, nanoscale and chemical character that enable unique capabilities across multiple fields; which are difficult to create using conventional methods. Thermally induced phase separation nanofibers, and electrospun nanofiber scaffolds, for example, are being developed and are widely regarded as an emerging technology and a potential strategy for biosensing, drug delivery, soft tissue regeneration, hard tissue regeneration, and wound healing. The capacity to alter numerous control aspects of the functional scaffold, such as fiber geometrical features and alignment, architecture, and subsequent material performance, is the technique's most prominent feature [1]. More importantly, electrospinning allows for the creation of a wide range of novel materials, including polymer alloys, nanoparticles, and active agents.

Nanofiber preparation employing the ES method has proved to be a future-proof materials technology, with numerous appealing characteristics such as outstanding mechanical properties and large specific surface areas. Due to the versatility, utility, and simplicity of the ES technology, the fibers produced are particularly appealing for numerous applications from a simple process capable of producing diverse morphologies [3]. The use of metal organic frameworks (MOFs) due to its flexible and functionalized molecular structures, nanofibers composites were fabricated as a novel molecular system with highly engineered structures for tailored applications. The usage of MOFs/carbon nanofibers (CNFs) as good electrode materials in energy transformation and storage technologies that include supercapacitors, sensors, and electrocatalysts is one of the most basic applications [4].

Electrospinning for materials technology of the future have seen a wide range of innovations of the technology including home-made re-designing of the technology to improve the ES apparatus reproducibility. Thus hybrid electrospun structures on different types of polymers have been developed and optimized to create products for various applications [5]. This chapter explores electrospinning innovation technology and the materials of the future, their properties and characteristics and applications. The focus materials of the future will be products fabricated from recyclable waste biomass materials as a way of valorization for higher hierarchical bioeconomic products.

2. Advances in electrospinning technology

The ability to tailor structural and morphological aspects of electrospun materials, such as the surface topography of nanofibers, and their porosity that allows enhanced mimicking of the manufactured material matrix, has sparked interest in the ES technology. This is accomplished by the ability to modify the electrospinning assembly in numerous ways in order to combine polymers with a wide range of materials (incorporate active materials such as drugs, inorganic catalysts, growth factors, functional groups and DNA/RNA as necessary in the various applications of the fabricated nanofibers [6]. **Figure 1** is a schematic diagram showing a simple set up of the electrospinning system.

Mokhtari et al. compared the technical assembly of the electrospraying and the electrospinning systems. This is because the two systems have different mechanisms of performing the fabrication of carbon materials they produce in unique distinct ways as shown in **Figure 2**. Electrospinning (**Figure 2a**) supports the formation of micro-scale and in some cases nano-scale fibers while the formation of thin films is facilitated by the electrospraying system (**Figure 2b**) aids in the formation of thin films [7]. As a result of insufficient polymer-chain entanglements in the polymer chains network, it was discovered that applying a high voltage below the minimum concentration causes electrospraying rather than electrospinning.

It was observed that varying the ratio of the polymer solution and the electrospinning potential difference results in the formation of unique materials ranging from beaded carbon deposits, heterogeneous fibers, uniform fibers, and entangled fibers [7]. Advanced efforts to improve electrospinning performance and the quality of the nanofibers while increasing cost-effective productivity of electrospinning and other nanofiber assembly technologies include integration of key concepts of conventional fiber production methods with nanotechnology. Electro-blowing, gas-jet/gas-assisted

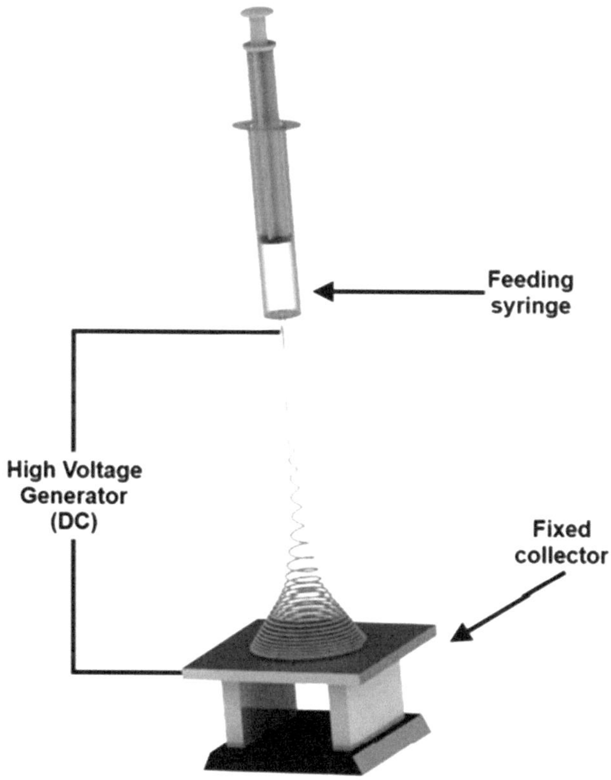

Figure 1.
Schematic illustration of vertical electrospinning setup [6].

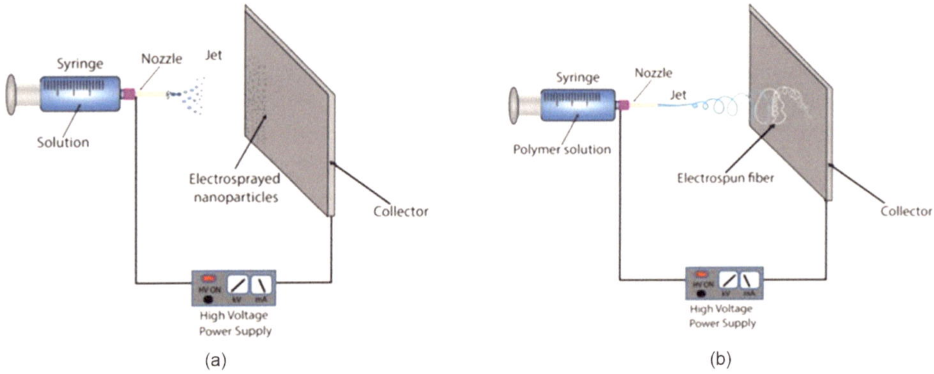

Figure 2.
(a) Schematic drawing of a typical electrospray setup. (b) Schematic drawing of a typical electrospinning setup.

electrospinning, and solution blowing, which advanced from melt blowing, combined with electro-centrifugal processing, centrifugal spinning, near field electrospinning with dip-pen nanolithography, and XanoShear, which combines shearing with wet spinning, are among the merged electrospinning conceptual technologies [8].

A look into a study of electrospinning as a versatile technique for fibrous material manufacturing in advanced fabrication of the electrospun biopolymer-based biomaterials compared the conventional needle-based and an innovative needless-based electrospinning processes. **Figure 3** presents the unique feature of the needless-based ES process is that the polymer solution is positioned in a bath and a high voltage polarized spinning mandrill is immersed into the bath.

When the rotating mandrill comes into contact with the grounded collection electrode, it collects a thin layer of polymer solution, which is subsequently subjected to an electric field. The electrostatic forces of the field at the needle's tip, or the thin layer of polymer solution at the rotating mandrill, overcome the solution's surface tension, pushing it to form several or a single Taylor cone, as illustrated in **Figure 4**. On its way to the collector, the charged polymer jet from the cones is ejected and extended.

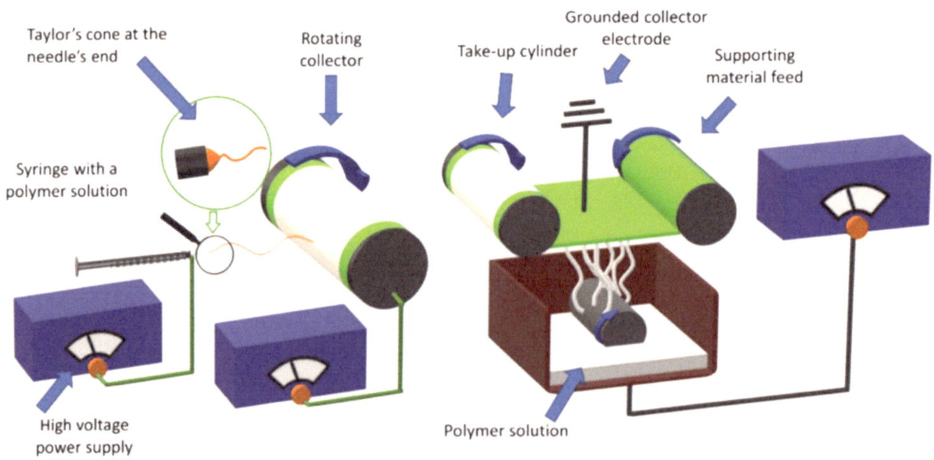

Figure 3.
Electrospinning setups needle-based (left) and needleless (right) [2].

Figure 4.
Needleless roller for electrospinning of polymer solutions.

The solvent evaporates from the solution, weakening the continuous jet of pure polymer and depositing it in a fibrous form on the collector [2].

A needleless mechanism performs the electrospinning of the polymer solution from the surface of a revolving roller. The roller is partially immersed in a tank containing material to be electrospun, as shown in **Figure 4**. On the roller's surface, a layer of consistently new material is generated by a rotating roller. When compared to needle electrospinning, the technique produces a large number of Taylor cones on the roller's surface, resulting in the technology's industrial applicability in mass manufacture of nanofibers materials [9].

Complex fibrous nanostructures have been prepared through manipulation of many experimental parameters of a multifluid electrospinning process. This is an innovative shift from the traditional single-fluid blending electrospinning process. However, there are difficulties in using multifluid processes, such as compatibility concerns of set up parameters including fluids, rate of stock feed and average proportions, interfacial tensions, and electrospinning sustainability [10]. Mass production of nanofibers using electrospinning was determined through the development of the macromolecular ES principle. The molecular flow in the spinning process, as well as the molecular direction in nanofibers, can be tailored to advance the electronic, and physico-chemical properties of nanofibrous materials, which influence their applications, molecular orientation in nanofibers, and structural hierarchical significance [11]. Several recent methods were developed to manufacture nanofibers using macromolecular ES processes. For example, industrial yarn production processes were only applicable for solution electrospinning via the innovative conceptualized gas-assisted melt ES. (GAME) as shown in **Figure 5**.

The unique characteristic of the innovative technique is the observed occurrence that turbulent air applies a pulling force, subsequently leading to an increase in output and a 10% reduction in melt jet width, with an additional 20-fold thinning when the air jet temperature is increased [12].

Multi-temperature control electrospinning (MTCES) is a practical way to spin molten polymers on a submicron level fiber than the conventional molten/solution ES. The molten precursor polymer was treated to quad-heating regions in the proposed MTCES design: needle, nozzle, rotating area, and collector to augment and regulate fiber size and morphology. The nozzle, spinning thermal parameters and dimensions, electric field, and flow rate of the MTCES are all adjusted to change the fiber diameter [13]. The MTCES setup is depicted in **Figure 6**. The technical mechanism demonstrates that the jet propagation begins to bend significantly near the collector at 25°C, and at 80°C, a strong melt jet propagation increases the dwelling time of the jet in the rotary region, demonstrating a distinct multi-control ES scheme, which was characterized by extensive preliminary work and models that used the same or similar setup schemes.

Energy materials have been fabricated by ES techniques as an alternative to fossil fuels and environmental mitigation initiatives. The nanofibrous materials produced by ES are extensively used in electrochemical energy storage devices. This is because the materials have inherent excellent properties, including an increased surface area, high dimensional ratio, good flexibility, high permeability, with several functionalities. A shift from the conventional ES methods saw the development of innovative enhanced ES techniques that produce nanofibers with novel special hierarchical nanostructures [14].

The core-shell structure was chosen because of its distinctive features, which can help to improve the preferred properties. Co-electrospinning creates core-shell fibers by filling two distinct precursor solutions into the double nozzles, as shown in **Figure 7** [14].

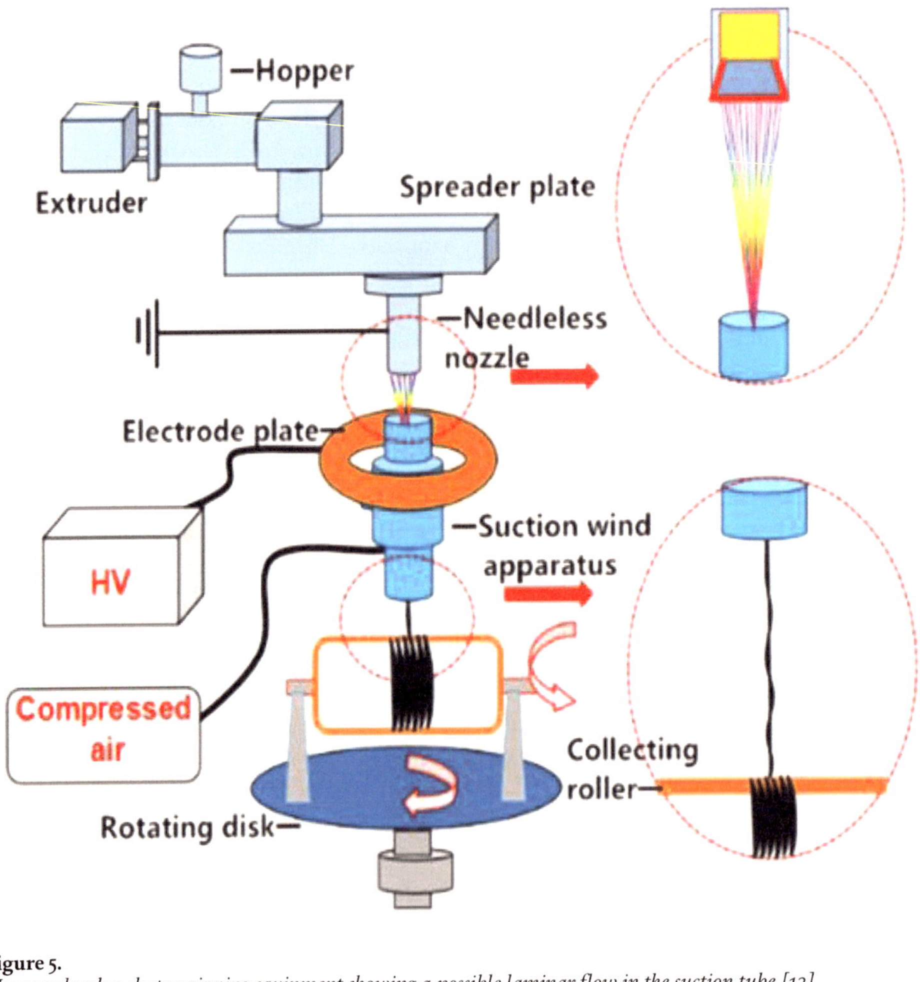

Figure 5.
Macromolecular electrospinning equipment showing a possible laminar flow in the suction tube [12].

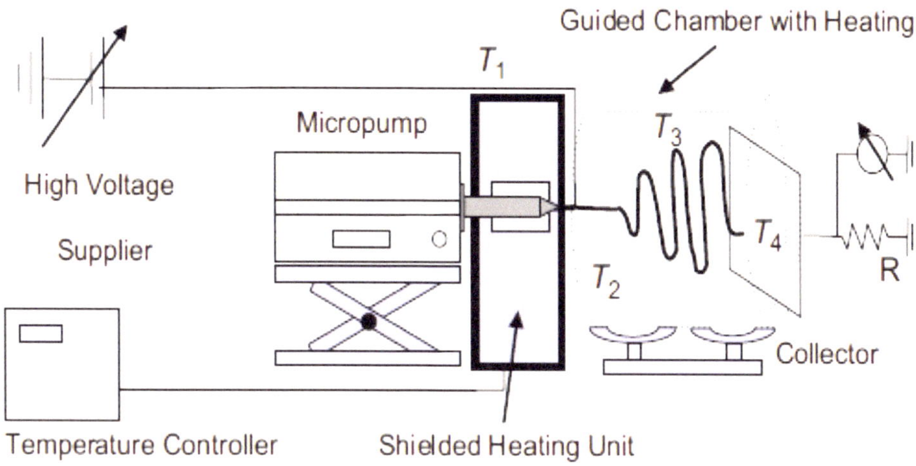

Figure 6.
The multi-temperature control electrospinning setup showing the multi-heating zone melt electrospinning [13].

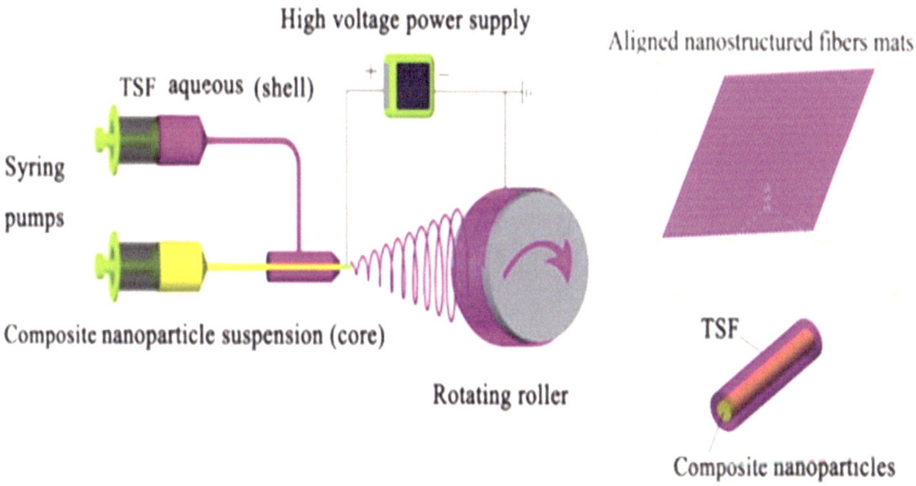

Figure 7.
Scheme for coaxial electrospinning [15].

The simplicity of setup and low cost, together with the ability to fabricate nanofibers with a wide range of compositions and morphologies, has aided ES technology's innovative advancement. Electrospinning-created nanofibrous structures provide appealing extracellular matrix conditions for the fixing, migration, and variation of materials matrix, including those giving rise to hard structure regeneration. The creation of structural materials regenerating nanofibers has been utilized by ES technology developments, which include material simulating composite/hybrid configurations and surface functionalization such as mineralization [16].

A special trifluid electrospinning technology was also developed as an innovation to the co-electrospinning process. This advancement provided for complex multi-chamber nanostructures for designing novel functional nanomaterials. The complex structure consisted of a collective shell and two independent openings of a multi-chamber nanostructure, with each having its own unique complex property, and these compartments form a total composite assembly within a region limited by nanofiber diameter. The sheath-separate-core fused nanostructure synchronized the functionalities of the three ES monolithic nanocomposites to afford a smart regulated release profile of a multi-chamber nanostructure, with each chamber characterized by distinct intrinsic complex property, and the structural compartments constituting a whole fused structure inside a section restricted by the diameter of nanofiber as shown in **Figure 8** [17].

Precision electrospinning, enabled by recent improvements in ES technology, is being envisioned as a viable option for fabricating 3D nanofibrous materials with a desired microstructure. Internal access to setup parameters such as solvent and fiber collecting method has increased intrinsic control of final nanofibrous architecture creation mechanism, as shown in **Figure 9** [18].

Plastic and other waste materials from industrial, domestic and agricultural activities, are the modern scourge on the face of the planet. The global call for re-use and recycle is gaining tremendous recognition with scientist scrambling for innovative ways of using waste materials in the circular economy. Waste biomass has been explored as an alternative source of polymers that may be used in wide range of ES processes targeting specific valorized products. As new materials use emerge and novel materials are electrospun into nanofibers, it is becoming increasingly critical to grasp current breakthroughs in biomass conversion into polymer sources for nanofibrous structures in order to fully exploit their potential. Advancements in waste

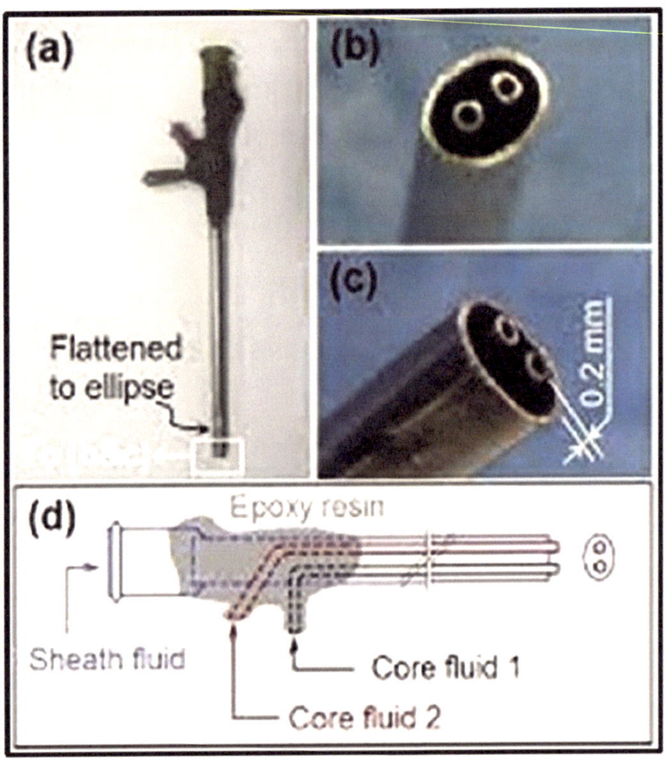

Figure 8.
Designs of the complex spinneret for implementing trifluid electrospinning: (a) a digital image showing a full view of the spinneret; (b) front view; (c) side view; and (d) a diagram about the organization of a structural outlet from three inlets [17].

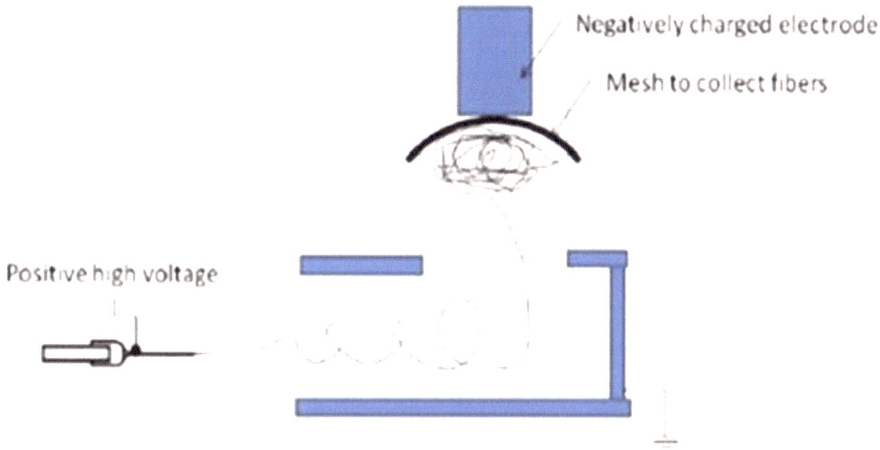

Figure 9.
Setup used to form 3D nanofibrous scaffold using a negatively charged electrode or negative ion generator [18].

biomass conversion technologies such as bio digestion, pyrolysis of plastic, and waste agricultural plant biomass wastes into bio oils and other polymers have preceded this.

3. Waste biomass feedstock for electrospinning nanofibers

Biomass is organic substances that is renewable and comprises plants and animals matter and may be combusted for heat or treated into renewable polymeric materials or fuels using a range of technologies. Most of the biomass end up as environmental waste materials that contaminate the land, rivers and oceans. Waste biomass include waste plant materials from crops, animal waste (dung and sewage), industrial waste in the form of effluent coming from industries such as petrochemical, food processing, textile dye effluent, pharmaceutical, and solid waste biomass including plastics, plant residues, (bagasse and other dregs), timber offcuts and sawdust, pulp and paper processing waste etc. These various biomass waste streams may be used as an alternative source of polymeric materials that may be used in electrospinning to produce materials for the future. Three classes of the waste biomass will be discussed namely synthetic waste biomass, natural flora waste biomass and natural fauna-based waste biomass.

3.1 Synthetic waste biomass

Plastic is the largest solid waste biomass on the face of the earth's surface while textile and pharmaceutical effluent are major synthetic liquid waste biomass. Unless great strides are made to valorize these waste streams and find hierarchical bioeconomic applications of these materials, they will persist in the environment as contaminants. Due to its tunable features, including wettability, surface charge, transparency, elasticity, porosity, and surface to volume proportion, various polymeric fibrous nano materials have been developed as simulated extracellular matrix. Using ES nanofibers of natural polymers (NPs) and synthetic polymers (SPs) as simulated extracellular matrix for tissue regeneration, a comprehensive investigation identified five basic kinds of nanofibrous polymers. NP–NP composites, NP–SP composites, SP–SP composites, cross-linked, and modified polymers with mineral materials are some of the polymers available [19]. Polycaprolactone (PCL), polylactic acid (PLA), poly(lactic-co-glycolic acid) (PLGA), and polyethylene terephthalate (PET) are some of prevalent well-known synthetic polymers [20, 21].

In recent years, a variety of processing technologies have been utilized in the manufacture of polymeric fibrous nano materials, including drawing, 3D printing, template synthesis, phase separation, self-assembly, ES, and so on. Synthetic ES nanofibrous materials processing allows for internal control of the electrospinning mechanism and foster chemical crosslinking to generate covalent connections between polymeric fibers. In either in situ electrospinning or post-spinning crosslinking, this manipulation is done to target qualities of the material of application in which the fibers will be used. Highly porous electrospun nanofibrous membranes, for example, have sparked a lot of interest in water filtering applications. **Figure 10** presents some of the common synthetic biomass materials used in ES of nanofibers. The creation of a reduced pore size and its distribution is highly favored by a thicker membrane with a lower mean fiber diameter, albeit the influence of membrane thickness is rather restricted. A high flux microfiltration (MF) sheath was fabricated based on efficient control of the total composite structure containing the electrospun layer thickness of 200 ± 10 m and a mean fiber diameter of 100 ± 20 nm [22].

A previous study looked at the spinnibility of various polymers, such as aqueous poly(ethylene oxide) (PEO) dispersed in alcohol-to-water mixtures. Fiber production was found to be possible with viscosities ranging from 1 to 20 poises and superficial tensions

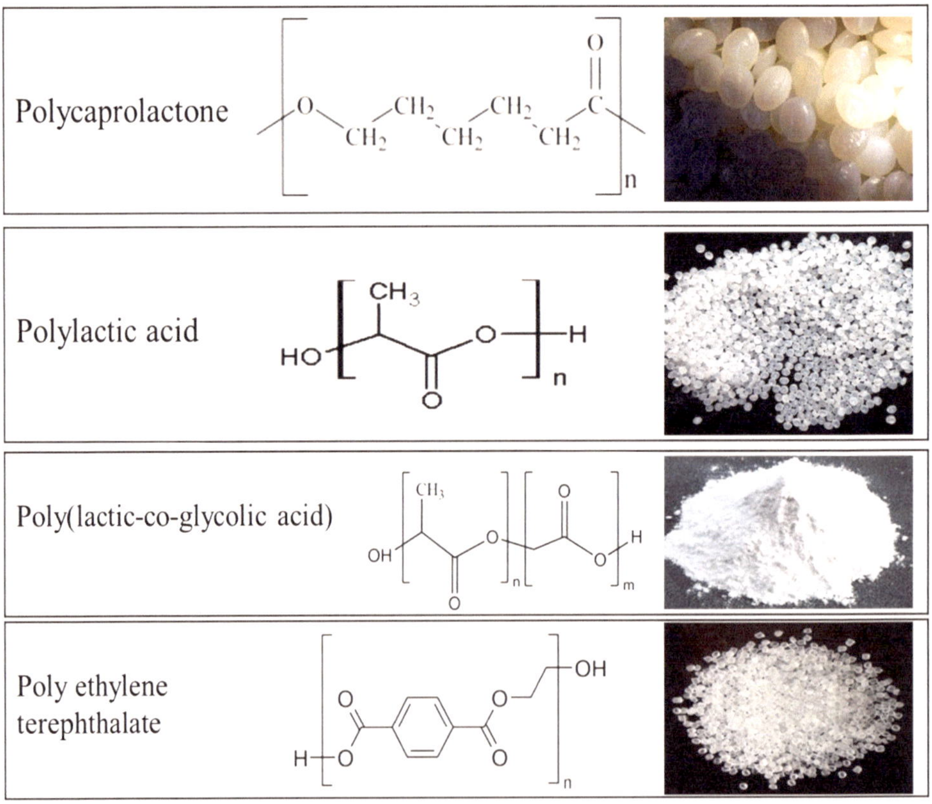

Figure 10.
Synthetic polymers used in electrospinning of nanofibrous materials.

of 35–55 dynes/cm. Electrospinning, however, was not feasible at viscosities more than 20 poises due to flow instability produced by the solution's high cohesiveness [23].

3.1.1 Natural flora waste biomass for electrospinning: material technology of the future

Spongy pomelo peels, rice husk, rice straw, sugar cane bagasse, coffee beans, coconut shells, and peanut shells have all been investigated as alternative sources of carbonaceous materials from biomass. In comparison to other carbonaceous precursors, these and other natural plant/floral biomass resources have grown increasingly appealing due to their availability, low cost, easy accessibility, and environmental friendliness. As a result, floral biomass has gotten a lot of attention in the electrospinning, biomedical, and energy storage fields [24]. Okara, soy pulp, or tofu dregs, for example, is a pulp made up of insoluble components of the soybean that remain after pureed soybeans are filtered for soy milk and tofu manufacture. Recent reviews have reported on the feasibility of ES fibrous nano materials made from a variety of decomposable and biocompatible matter, including natural proteins like floral and faunal collagen, gelatin, silk, chitosan, and alginate [25].

The preparation of the waste floral/plant biomass for ES of nanofibrous materials involves a number of steps that extract plant proteins in the insoluble parts of the waste biomass. Silk fibroin (SF), for example, is made by degumming raw silk fibers twice with a 0.5% (W/W) $NaHCO_3$ base medium at 100°C, over half an hour period followed by

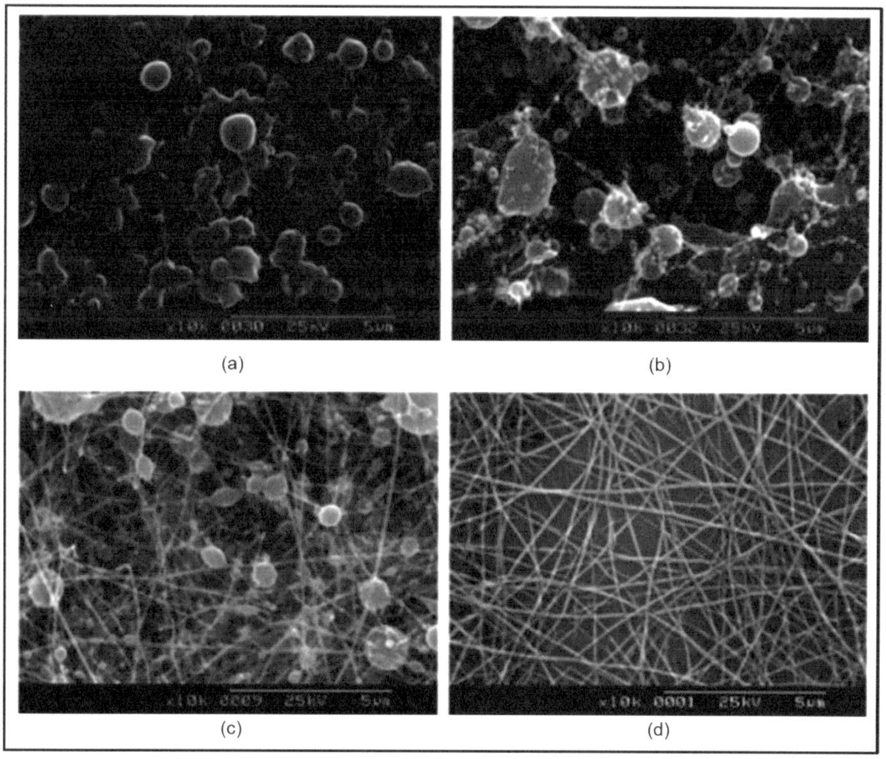

Figure 11.
SEM micrographs of electrospun SF nanofibers with concentration of (a) 3%, (b) 6%, (c) 9%, and (d) 12% [26].

rinsing with warm dH_2O. At 70°C for 6 h, degummed silk (SF) is dispersed in a ternary aqueous medium of calcium chloride-ethanol-water (1:2:8 in molar ratio). The SF was filtered and lyophilized after 3 days of dialysis using cellulose hollow sheath (250-7u; Sigma) in dH_2O to get the regenerated SF sponges. Dispersing the SF sponges in 98% methanoic acid (Aldrich) for 3 h makes SF solutions. The molar quantities of SF solutions for electrospinning range from 3% to 15% by weight [26].

Extracted silk fibroin was used to prepare silk electrospinning as presented in **Figure 11**. Electrospun SF nanofibers with varied silicon fibroin concentrations of 3%, 6%, 9%, and 12% are shown in SEM micrographs. The most prevalent natural polymers used as ES nanofiber materials include chitosan, collagen, gelatin and silk [20, 21]. Natural polymer nanofibers present distinguished features like biodegradability and biocompatibility, a phenomenon that makes them suitable materials in biological environments. **Figure 12** presents some of the abundant natural polymers adapted for ES nanofibers production. Chitin and its over 50% deacetylated derivative, chitosan, for example, are commonly used natural polysaccharides as scaffolds. Blending with other materials are thus required to tailor-make materials with a set of acceptable features and attributes in order to achieve a stronger composite. Chitin/silk fibroin (chitin/SF) nanofibers, for example, were used to make novel ECM scaffolds [27].

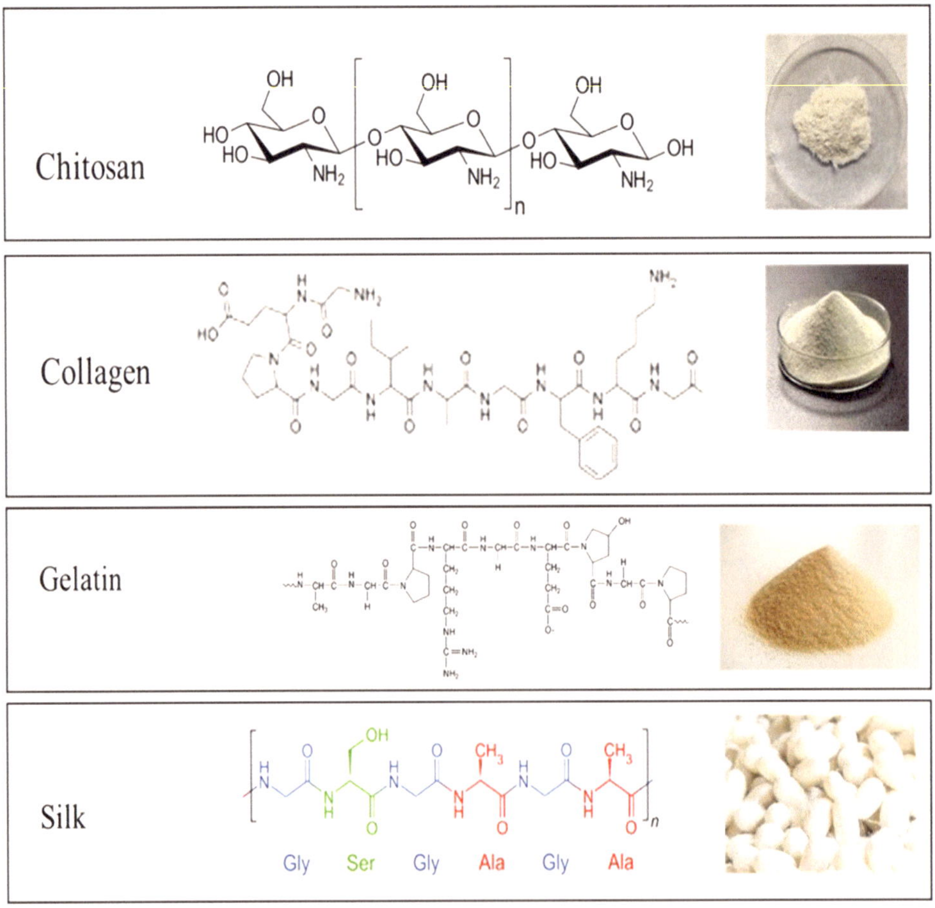

Figure 12.
Natural polymers used in electrospinning of nanofibrous materials.

Biocompatibility and biological activity are two characteristics of natural polymers. However, these polymers have some drawbacks, such as engineering and processing difficulties due to poor mechanical strength, restricted processing and manufacturing capacities, batch-to-batch variability, and the possibility of pathogen transmission [20]. Collagen and proteoglycans, for example, make up the majority of the body's natural extracellular matrices (ECMs), which vary in composition depending on tissue type. Nanofibrous scaffolds made of collagen fused with glycosaminoglycans (GAGs), the major constituent of proteoglycans like condroitin sulfates and hyaluronic acid, are suitable for creating a perfect scaffold that mimics the natural ECM. Collagen and GAGs' utility, on the other hand, has been limited because of their exorbitant price and poor mechanical qualities. In biomedical applications, this phenomenon can be addressed by fusing natural polymers such as proteins polymeric strands and polysaccharides fibrous materials, which can improve biotic transformation of cells and accelerate tissue development [27].

Most of the insoluble floral biomass is in the form of lignocellulosic and chitin material. The success of tapping into the floral biomass as a resource for ES of nanofibrous material depends on the ability to depolymerize the lignin and chitin long polymer chains. It is these polymers that will be used for ES processes to produce electrospun nanofibers. Recently, there has been renewed interest in producing carbon fibers from sustainable cellulosic precursors [28]. The abundance and cost effectiveness of cellulose as a material generator, as well as the relatively ecologically friendly fiber production methods used preceded this interest. Recent research on regenerated cellulose fibers from a fluid crystalline fabrication route as a carbon fiber precursor generated strands with a modulus of 140 GPa for the shell area and 40 GPa for the core area, indicating that CNFs resulting from nano-sized cellulosic precursors are even more competent as physical reinforcement than micron-sized fibers; because of their reduced diameters, providing a greater surface area for bonding and stress transfer [29].

3.1.2 Natural fauna waste biomass for electrospinning: material technology of the future

Animal manure, agricultural residues, organic portion of municipal solid garbage, industrial waste biomass, and natural vegetation cycle waste are all examples of enormous amounts of organic waste produced by many sectors. Similarly, fauna waste biomass, primarily in the form of keratin, a durable, fibrous protein found in advanced vertebrates (mammals, birds, and reptiles) and human epithelial cells, has been widely employed in ES for the creation of nanofibrous materials. Millions of tons of keratin-containing biomass are produced by the food business, particularly the meat market, slaughterhouses, and wool manufacturers. These sectors are rapidly expanding, with the United States, Brazil, and China accounting for more than 40 million tons of fauna-based biomass annually [30]. Inadequate management of these organic wastes can harm the environment by polluting water and air, lowering people's quality of life [31].

If controlled with scientific interventions, organic waste no longer persists as garbage, but instead becomes a rich source of substrate, polymers, and molecules for the production of a variety of value ES nanofibrous products [32]. Detailed studies explored potential applications of the fauna generated organic waste in the production of biogas for energy production. Human waste is disposed of as sewage in the form of biological wastewater. Technological advances unravelled biological wastewater treatment plants (WWTP) as an approach to converting biomass into rich materials for precursor molecules for polymerization in ES nanofibrous material fabrication or for energy production [33]. Fauna waste biomass in the form of dung (**Figure 13**), piggery or fowl wastewater treatment with purple phototrophic bacteria was explored as a promising platform for electrospinning biomass resource recovery process under optimized operational conditions [34].

Figure 13.
Fauna biomass: cow dung is co-digested with sewage for production of gas in an anaerobic bio digester.

Figure 14.
Sewage treatment plant for gas production.

It is important to note that fauna waste biomass is a natural phenomenal bio digestive process of converting lignocellulosic and chitin organic biomass and transform it into shorter chains of polysaccharides and other polymeric substrates for ES nanofibrous materials production. Anaerobic bio digestion followed by catalytic polymerization of biogas molecules such as methane, ethane and propane, will produce tailor-made polymeric materials that may be used in electrospinning production of carbon nanofibrous materials. **Figure 14** is an advanced industrial scale bio digestion plant for production of biogas.

Bio digestion of fauna waste biomass is a significant alternative supply of materials for electrospinning of nanofibrous materials when modern methods are used. Previous research on bio digestion of fauna waste biomass for methane production found that the influence of pre-treatment results in a substantial increase in gas production of up to 67%, with a 52% methane content in the biogas. As a result, it was determined that pretreatment of both feed and biomass improves biogas output but not methane content [35]. According to recent studies, the valorization of bio or organic waste is being prioritized in order to tackle the rapid accumulation of waste generated from food production activities, as well as to create sustainable feedstock for industrial materials and chemicals in place of fossils and synthetic materials (see Section 3.1). Biogas, compost, and small platform molecules are currently produced from biowaste via anaerobic bio digestion, fermentation, and thermo-chemical methods as shown in **Figure 15**. There are currently no commercial low-temperature chemical methods for valorizing organic lignin fractions as feedstock for modified compounds. Thus, research has been conducted to fill this technological gap, demonstrating that moderate thermal hydrolysis of municipal bio-waste manure reserve is a safe, environmentally sustainable, and affordable process for transforming lignin-like material from compost into value-added specialty chemicals for the production of ES nanofibrous materials (**Figure 15**) [37].

Biomass is a readily available and long-lasting ES material that may be turned into carbon based smart energy storage device and other uses. For carbon nanofiber manufacture, many strategies were used to meet various goals, including an increased productivity, easy dimensional parameters manipulation, energy efficient, and a high turnover. Nonetheless, several critical features of biomass-based fibrous carbon nano materials

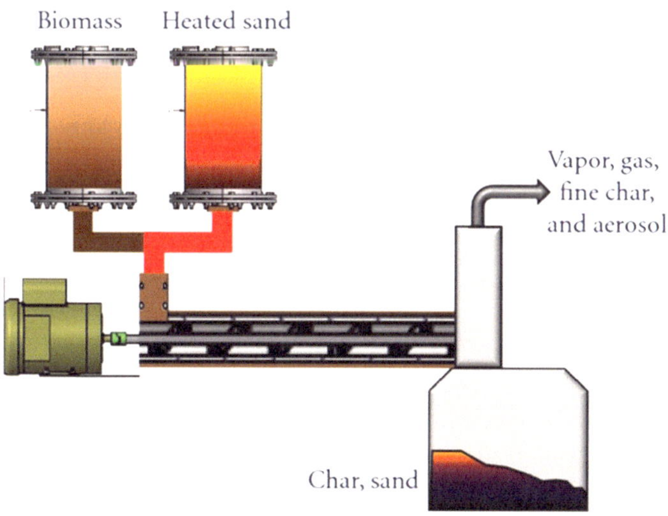

Figure 15.
Auger/screw pyrolysis reactor concept using heat carrier [36].

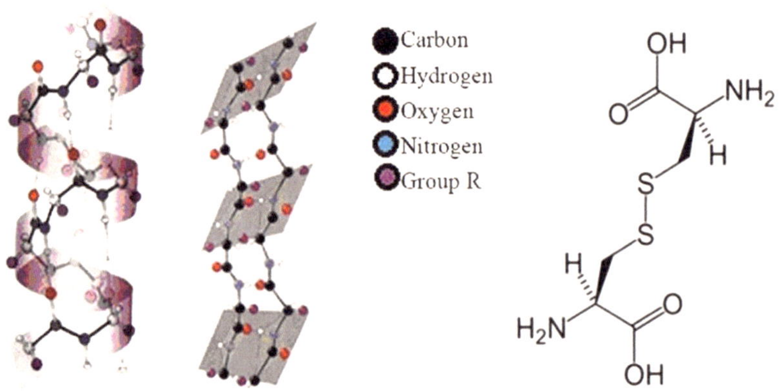

Figure 16.
An α-helix and β-pleated sheet keratin and the molecular structure [40].

are yet to be extensively studied, thus information gaps still exist for each process to be supplied. As a result, more research is needed to expand our knowledge of the essential characteristics of various processes in order to generate highly desirable precursor materials for ES fibrous carbon nano materials manufacture from organic matter for sustainable materials manufacturing and energy smart storage applications [38].

An example of fauna waste biomass material rich in extractable materials for ES nanofibers materials is feathers from the poultry industry. Chicken feathers, comprises 90% raw keratin protein and 70% amino acids, can be employed as one of the primary sources for extracting keratin. Keratin is used in a variety of industries, including biotechnology, waste management, cosmetics, and medicine [39]. Waste feathers can be converted into keratin in a cost-effective and environmentally beneficial manner. Keratin is an insoluble protein of the cytoskeletal element with a size of 8–10 nm that belongs to a group known as intermediate filaments (IFs). Keratin is a fibrous protein with a helical structure, as seen in **Figure 16**, and is the ecosystem's third most prevalent natural biomass polymer after chitin and cellulose [41].

4. Innovative waste biomass-sourced electrospun products

Electrospun fibers fabricated from waste biomass sources has resulted in manufacturability of bioactive electrospun nanofibers and has been reported as potential drug delivery agents [42], wound dressing with antibacterial activity, filtration, cosmetics, protective clothing, electrical applications [43] catalysis [44], food industry [44], facial mask [45], and smart energy storage devices, such as supercapacitors as illustrated in **Figure 17**.

Natural biopolymer electrospun products are made up of ultrafine fibers that are reusable, nontoxic, biocompatible, biodegradable and antibacterial properties. The fibers have been reported to possess excellent physical and chemical characteristics such as high degrees of crystallinity, aspect ratio, large specific surface area, number of surface hydroxyl groups, thermal resistance and excellent mechanical properties [45, 46]. However, the substantial chemical and energy consumption associated with the isolation of macro-sized fibers to nano-sized fibers creates manufacturing hurdles for waste bioactive electrospun nanofibers [46]. As a result, findings on waste bioactive electrospun nanofibers are still in their infancy in the literature [46].

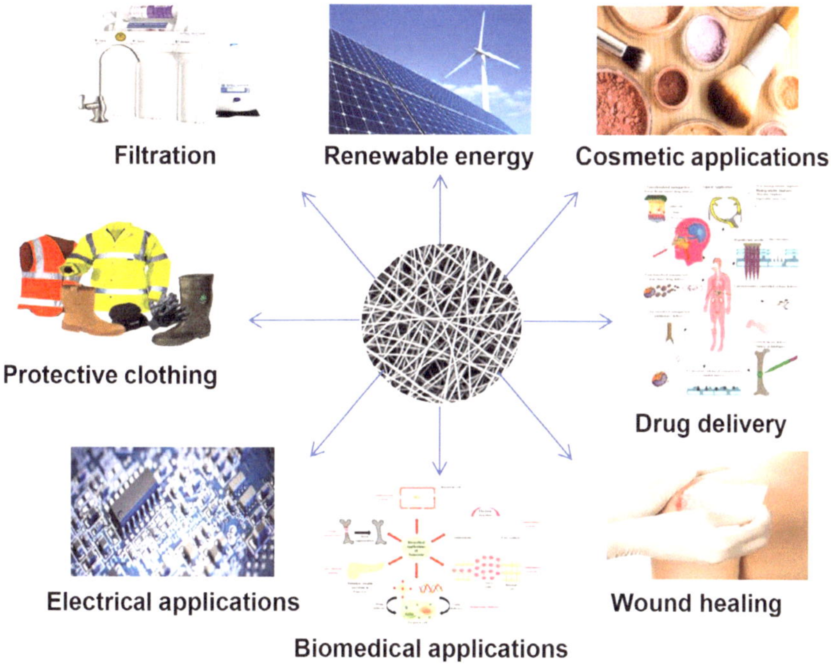

Figure 17.
Applications of nanofibers in different fields for day to day activities.

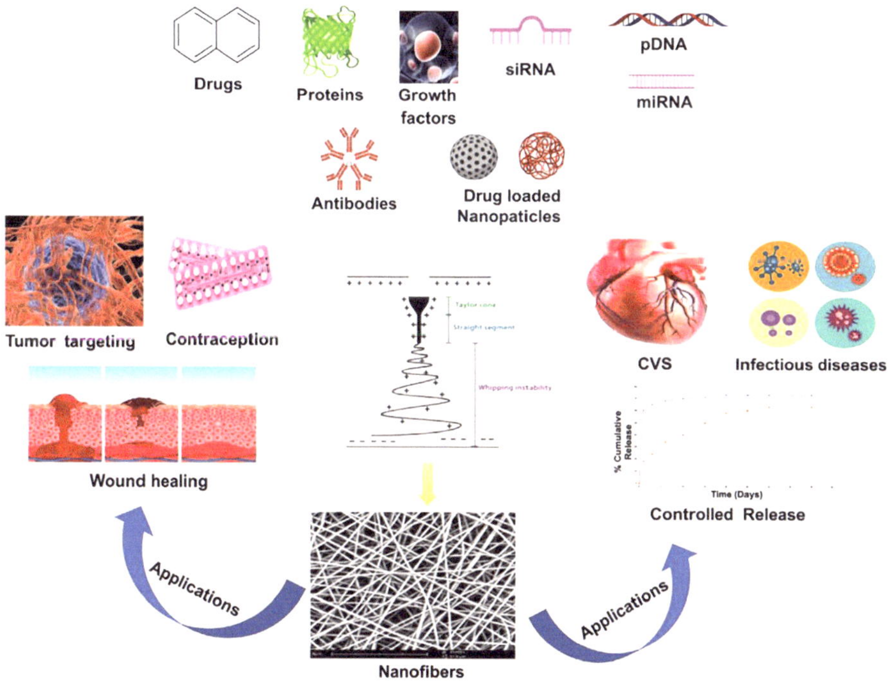

Figure 18.
Illustration of various applications of bioactive electrospun fibers in the biomedical field [43].

4.1 Biomedical product

In the biomedical field, literature reports on manufactured products made from biomass electrospun fibers range from medication delivery agents to biomaterials [42], wound dressing with antibacterial activity, facial mask [45], and tissue regenerative biomedical applications as presented in **Figure 18**.

The ultrafine fibers have been previously reported to result in high-performance filters and applicability in facial masks [45, 47]. Various ultrafine fiber filters have been created that can filter particles larger than 10 nm with excellent efficiency. Spider-web network filters are described in the literature as having a combination of extremely efficient, long-range electrostatic property, low air resistance, and great transparency [45, 47]. Viruses can be blocked by ultrafine fiber filters [47]. Irrespective of the challenges associated with the fabrication of bioactive electrospun fibers products. The choice of polymer used aid in fabricating fibers with antimicrobial activities [45].

Figure 19a presents a typical electrospinning technology. Choice of polymer, concentration, flow rate, needle, and tip-to-collector distance all affect fiber quality. **Figure 19b** shows various types of electrospun fibers. The structure of a hybrid filter that works as both a filter and a hydrophobic layer is shown in **Figure 19c** and **d**.

Facial masks constructed from electrospun biomass possess key characteristic performance features that has the potential to outcompete with the masks in the market. Advantages of biomass electrospun masks vary from the transparent, reusability, antiviral, degradable smart masks that possess filtration, thermal stability, and water resistance [45]. The facial mask technique has a wide range of possible uses, including filtration systems in water treatment, protective garments, and cosmetics [45].

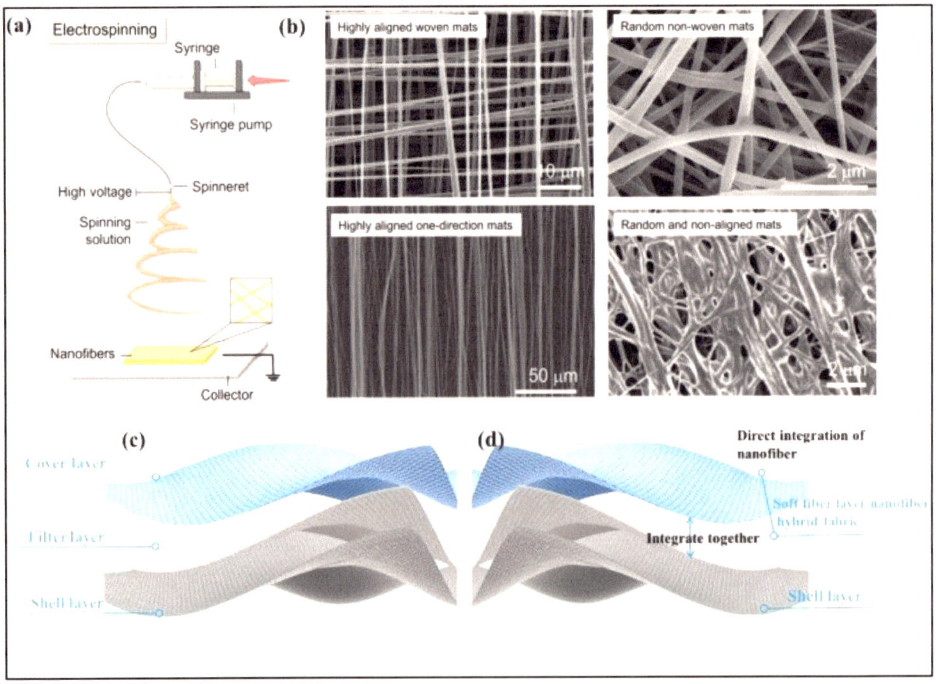

Figure 19.
(a) Scheme of electrospinning technology. (b) Various SEM images of electrospun nanofibers. (c) Scheme of generally utilized masks. (d) The proposed structure of electrospun ultrafine fibrous masks.

As a result of the structure and bioactivity of loaded pharmaceuticals remaining unaltered during the spinning process, electrospun drug-delivery agents drew interest. They also reduced in vitro drug burst release and can contain a range of biomolecules [48]. Drug delivery agents fabricated from all forms of cellulose polymer results in drug delivery systems that are hydrophilic, eco-friendly, bio-degradable, and biocompatible [42].

4.2 Renewable energy products

Incorporation of NPs, natural biomass onto the polymer through the electrostatic interaction between their functional groups has a stabilizing effect on NPs [44, 49]. These electrospun catalyst found application in catalysis, supercapacitors, corrosion inhibition, and within the food industry natural polymers [44, 49].

Carbon-based supercapacitors with a large interactive surface and high permeability have sparked interest in natural floral and faunal waste materials, owing to the growing ecological consciousness. Electrospun cellulose-based supercapacitors are still in the laboratory stage, despite their rich carbon abundance of roughly 44%, great stability, and exceptional permeability linked with its hierarchical conformation and exceedingly efficient rigid lateral chains in cellulose [49]. The energy density of cellulose-based supercapacitors is low [49]. Hence the poor electrical performance and cell voltage. Another limitation is time consumption associated with economic factor in the optimization stage of cellulose electrospun mats.

As an alternate technique for increasing the electrochemical properties of lignin/cellulose nanofiber electrodes, creating compound electrode materials with a lignin/cellulose backbone can be used to address these constraints [49, 50]. Literature presented flexibility, wide surface area, outstanding mechanical flexibility, and particularly good electrical conductivity, composite nanofibers and ES activated carbon fiber network (ACFN) as attributes to improved performance. When employed as supercapacitor electrodes, they have a high electrical performance, a phenomenon attributed to their pseudo-capacitance [51, 52]. As a result, ACFNs lignin/cellulose nanofiber composites could be an attractive electrode material for biomass-based flexible supercapacitors [49]. Furthermore, when the electrolyte penetrates the micropores of the electrospun mats, as shown in **Figure 20**, the characteristics of the electrospun biomass composites can be adjusted, allowing for the wettability feasible with the preferred electrolyte [53].

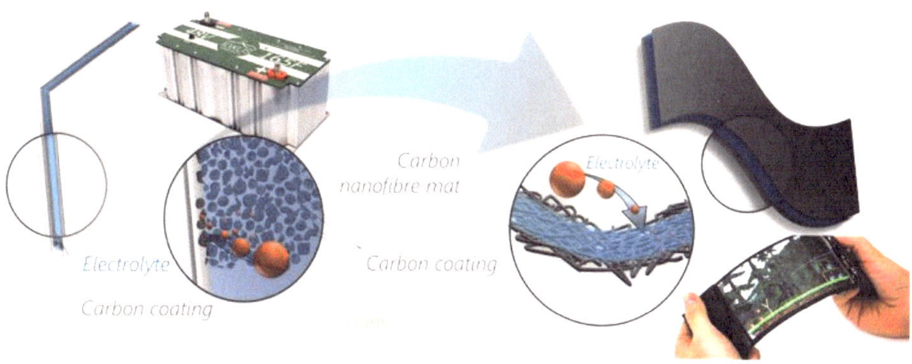

Figure 20.
Supercapacitive cell with thin film-coated carbon powder-based electrodes and free-standing and flexible flexible carbon nanofiber electrodes in conjunction with a polymer electrolyte [50].

In aqueous electrolytes, heteroatoms have been reported to enhance wettability of carbonaceous surfaces [54]. Lignin has a lot of oxygen functional groups and a lot of active hydrophilic surface. However, biomass-derived ECNF p-doped performed worse relative to the commercial CF. The lower performance could be attributable to the starting material's higher number of oxygen functional groups. P-doping has been reported to block micro/mesopores, reduce conductivity and electron transport [50]. Jet viscosity of the polymer was not measured, as such further research still has to be done.

As a result, environmentally friendly biomass electrospun fibers with improved performance in working electrochemical devices have demonstrated that the fabrication of future smart energy storage materials will be ecologically viable, providing a completely green alternative to the powering of transportation and conventional storage [50].

4.3 Electrical products

The versatility of waste biomass electrospun fibers, as well as their controllable physical and chemical properties, make them a model technique for electrode fabricating and flow media for a variable of smart energy devices, with the ability to reduce mass transport and activate overpotentials, thereby increasing competence [50]. Natural biomass is being used as a polymer of choice because of its capacity to infuse sustainable principles in electrochemical device materials. This also contributes to their capacity to increase the use of renewable electricity through their application [50]. Lignin is a waste by-product derived from natural flora that has been documented to exist in three different types: Different molecular weights and mechanical and thermal stabilities of kraft (KL), ethanol organosolvents (EOL), and phosphoric acid lignin (PL) [50]. For vanadium redox couples, electrospun carbon nanofibers produced from PL and KL at 9 kV demonstrated excellent cyclic voltammetry electrochemical performance. **Figure 20** clearly illustrates potential electrical products that can be fabricated from waste biomass electrospun fibers. Redox flow batteries (RFBs), fuel cells, and metal air batteries are some of the potential products shown in **Figure 20** [50]. The use of electrospun material in RFBs is still in its infancy and requires further development. Nonetheless, the improved redox couple's catalytic activity of waste biomass electrospun fibers provides an alternate solution to commercial electrodes' high overpotential when discharge current density is large [50].

Electrospun fibers made from waste biomass have the potential to be used in redox flow batteries because they form microstructures with large surface areas and mass transport qualities in the electrodes. Similarly, improved biomass electrospun fiber applicability in fuel cells and metal air batteries offers a conductive-advanced structure for the gas diffusion layers that can dope and/or support catalytic nanoparticles, as well as electrochemically active fibers [50].

5. Conclusion and future works

The advancement of electrospinning (ES) technologies and the industrial production of ES fibrous carbon nano materials to suit or facilitate different bioeconomic uses was aided by technical innovation. It may be inferred that the capacity to change the electrospinning assembly in various ways, in order to combine different materials with a wide variety of properties as well as incorporate active elements, will have a substantial impact on the production of materials in the future. By combining essential concepts from traditional fiber manufacturing techniques with nanotechnology, the performance of electrospinning technology and the quality of nanofibers can be increased. In comparison to other carbonaceous precursors, natural flora and fauna waste biomass for future electrospinning material technology has become increasingly appealing due to its abundance, low cost, easy accessibility, and environmental friendliness. Most of the insoluble floral biomass is in the form of lignocellulosic and chitin material while the soluble biomass is in the form of proteins and polysaccharides. Fauna waste biomass is mainly in the form of keratin. Millions of tons of keratin biomass are produced by industry, particularly the meat market, slaughterhouses, and wool manufacturers. The determination of marketable low thermal chemical procedures to valorize bio and organic waste lignin fractions as feedstock for commercial chemicals will be the focus of future work aimed at advancing electrospinning materials.

Author details

Farai Dziike[1]*, Phylis Makurunje[2] and Refilwe Matshitse[3]

1 Technology Transfer and Innovation Directorate, Durban University of Technology, Durban, South Africa

2 Nuclear Futures Institute, Bangor University, Gwynedd, Wales, United Kingdom

3 Department of Chemistry, Rhodes University, Grahamstown, South Africa

*Address all correspondence to: faraid1@dut.ac.za

References

[1] Kchaou M, Alquraish M, Abuhasel K, Abdullah A, Ali AA. Electrospun nanofibrous scaffolds: Review of current progress in the properties and manufacturing process, and possible applications for COVID-19. Polymers (Basel). 2021;**13**:1-20

[2] Wilk S. Advances in fabricating the electrospun biopolymer-based biomaterials. Journal of Functional Biomaterials. 2021;**12**:1-32

[3] Wang C, Wang J, Zeng L, Qiao Z, Liu X, Liu H. Fabrication of electrospun polymer nanofibers with diverse morphologies. Molecules. 2019;**24**:4-33. DOI: 10.3390/molecules24050834

[4] Mini-review NEA, Liu M, Cai N, Chan V, Yu F. Development and applications of mofs derivative one-dimensional development and applications of mofs derivative one-dimensional nanofibers via electrospinning: A mini-review. Nanomaterials. 2019;**9**:1-21. DOI: 10.3390/nano9091306

[5] Dias JR, Granja PL, Bártolo PJ. Advances in electrospun skin substitutes. Progress in Materials Science. 2016;**84**:314-334

[6] Sasmazel H, Ozkan O. Advances in electrospinning of nanofibers and their biomedical applications. Current Tissue Engineering. 2013;**2**:91-108. DOI: 10.2174/22115420113029990007

[7] Mokhtari F, Salehi M, Zamani F, Hajiani F, Zeighami F, Latifi M, et al. Advances in electrospinning: The production and application of nanofibres and nanofibrous structures. Textile Progress. 2016;**48**:119-219. DOI: 10.1080/00405167.2016.1201934

[8] Luo CJ, Stoyanov SD, Stride E, Pelan E, Edirisinghe M. Electrospinning versus fibre production methods: From specifics to technological convergence. Chemical Society Reviews. 2012;**41**: 4708-4735. DOI: 10.1039/c2cs35083a

[9] Jirsak O, Petrik S. Recent advances in nanofibre technology: Needleless electrospinning. International Journal of Nanotechnology. 2012;**9**:836-845. DOI: 10.1504/IJNT.2012.046756

[10] Zhao K, Wang W, Yang Y, Wang K, Yu DG. From Taylor cone to solid nanofiber in tri-axial electrospinning: Size relationships. Results in Physics. 2019;**15**:4-6. DOI: 10.1016/j.rinp.2019.102770

[11] Tian D, He JH. Macromolecular electrospinning: Basic concept & preliminary experiment. Results in Physics. 2018;**11**:740-742. DOI: 10.1016/j.rinp.2018.10.042

[12] Ibrahim YS, Hussein EA, Zagho MM, Abdo GG, Elzatahry AA. Melt electrospinning designs for nanofiber fabrication for different applications. International Journal of Molecular Sciences. 2019;**20**:1-17. DOI: 10.3390/ijms20102455

[13] Zhou H, Green TB, Joo YL. The thermal effects on electrospinning of polylactic acid melts. Polymer (Guildf). 2006;**47**:7497-7505. DOI: 10.1016/j.polymer.2006.08.042

[14] Liu Q, Zhu J, Zhang L, Qiu Y. Recent advances in energy materials by electrospinning. Renewable and Sustainable Energy Reviews. 2018;**81**:1825-1858. DOI: 10.1016/j.rser.2017.05.281

[15] Shao W, He J, Sang F, Ding B, Chen L, Cui S, et al. Coaxial electrospun aligned tussah silk fibroin nanostructured fiber

scaffolds embedded with hydroxyapatite-tussah silk fibroin nanoparticles for bone tissue engineering. Materials Science and Engineering: C. 2016;**58**:342-351. DOI: 10.1016/j.msec.2015.08.046

[16] Shin SH, Purevdorj O, Castano O, Planell JA, Kim HW. A short review: Recent advances in electrospinning for bone tissue regeneration. Journal of Tissue Engineering. 2012;**3**:1-11. DOI: 10.1177/2041731412443530

[17] Chang S, Wang M, Zhang F, Liu Y, Liu X, Yu DG, et al. Sheath-separate-core nanocomposites fabricated using a trifluid electrospinning. Materials and Design. 2020;**192**:108782. DOI: 10.1016/j.matdes.2020.108782

[18] Teo WE, Inai R, Ramakrishna S. Technological advances in electrospinning of nanofibers. Science and Technology of Advanced Materials. 2011;**12**:1-19. DOI: 10.1088/1468-6996/12/1/013002

[19] Keshvardoostchokami M, Majidi SS, Huo P, Ramachandran R, Chen M, Liu B. Electrospun nanofibers of natural and synthetic polymers as artificial extracellular matrix for tissue engineering. Nanomaterials. 2021;**11**: 1-23. DOI: 10.3390/nano11010021

[20] He B, Zhao J, Ou Y, Jiang D. Biofunctionalized peptide nanofiber-based composite scaffolds for bone regeneration. Materials Science and Engineering: C. 2018;**90**:728-738. DOI: 10.1016/j.msec.2018.04.063

[21] Meng ZX, Wang YS, Ma C, Zheng W, Li L, Zheng YF. Electrospinning of PLGA/gelatin randomly-oriented and aligned nanofibers as potential scaffold in tissue engineering. Materials Science and Engineering: C. 2010;**30**:1204-1210. DOI: 10.1016/j.msec.2010.06.018

[22] Wang R, Liu Y, Li B, Hsiao BS, Chu B. Electrospun nanofibrous membranes for high flux microfiltration. Journal of Membrane Science. 2012;**392-393**:167-174. DOI: 10.1016/j.memsci.2011.12.019

[23] Huang ZM, Zhang YZ, Kotaki M, Ramakrishna S. A review on polymer nanofibers by electrospinning and their applications in nanocomposites. Composites Science and Technology. 2003;**63**:2223-2253. DOI: 10.1016/S0266-3538(03)00178-7

[24] Ru H, Xiang K, Zhou W, Zhu Y, Zhao XS, Chen H. Bean-dreg-derived carbon materials used as superior anode material for lithium-ion batteries. Electrochimica Acta. 2016;**222**:551-560. DOI: 10.1016/j.electacta.2016.10.202

[25] Kai D, Liow SS, Loh XJ. Biodegradable polymers for electrospinning: Towards biomedical applications. Materials Science and Engineering: C. 2015;**45**:659-670. DOI: 10.1016/j.msec.2014.04.051

[26] Min BM, Lee G, Kim SH, Nam YS, Lee TS, Park WH. Electrospinning of silk fibroin nanofibers and its effect on the adhesion and spreading of normal human keratinocytes and fibroblasts in vitro. Biomaterials. 2004;**25**:1289-1297. DOI: 10.1016/j.biomaterials.2003.08.045

[27] Park KE, Jung SY, Lee SJ, Min BM, Park WH. Biomimetic nanofibrous scaffolds: Preparation and characterization of chitin/silk fibroin blend nanofibers. International Journal of Biological Macromolecules. 2006;**38**:165-173. DOI: 10.1016/j.ijbiomac.2006.03.003

[28] Dumanlı AG, Windle AH. Carbon fibres from cellulosic precursors: A review. Journal of Materials Science. 2012;**47**:4236-4250. DOI: 10.1007/s10853-011-6081-8

[29] Deng L, Young RJ, Kinloch IA, Zhu Y, Eichhorn SJ. Carbon nanofibres

produced from electrospun cellulose nanofibres. Carbon N. Y. 2013;**58**:66-75. DOI: 10.1016/j.carbon.2013.02.032

[30] Sharma S, Gupta A. Brazilian archives of biology and technology sustainable management of Keratin Waste Biomass: Applications and future perspectives. Brazilian Archives of Biology and Technology. 2016;**5959**:1-14. DOI: 10.1590/1678-4324-2016150684

[31] Sarikaya E, Demirer GN. Biogas production from broiler manure, wastewater treatment plant sludge, and greenhouse waste by anaerobic co-digestion. Journal of Renewable and Sustainable Energy. 2013;**5**:1-12. DOI: 10.1063/1.4818771

[32] Bhatia RK, Ramadoss G, Jain AK, Dhiman RK, Bhatia SK, Bhatt AK. Conversion of waste biomass into gaseous fuel: Present status and challenges in India. Bioenergy Research. 2020;**13**:1046-1068. DOI: 10.1007/s12155-020-10137-4

[33] Bodík I, Bodík I, Sedláèek S, Kubaská M, Hutòan M. Biogas production in municipal wastewater treatment plants-current status in EU with a focus on the Slovak Republic. Chemical and Biochemical Engineering. 2011;**25**:335-340. Available from: https://www.researchgate.net/publication/232594316

[34] Sepúlveda-Muñoz CA, de Godos I, Puyol D, Muñoz R. A systematic optimization of piggery wastewater treatment with purple phototrophic bacteria. Chemosphere. 2020;**253**:1-9. DOI: 10.1016/j.chemosphere.2020.126621

[35] Divyalakshmi P, Murugan D, Sivarajan M, Sivasamy A, Saravanan P, Rai CL. Effect of ultrasonic pretreatment on secondary sludge and anaerobic biomass to enhance biogas production. Journal of Material Cycles and Waste Management. 2018;**20**:481-488. DOI: 10.1007/s10163-017-0603-7

[36] Verma M, Godbout S, Brar SK, Solomatnikova O, Lemay SP, Larouche JP. Biofuels production from biomass by thermochemical conversion technologies. International Journal of Chemical Engineering. 2012;**12**:1-18. DOI: 10.1155/2012/542426

[37] Negre M, Montoneri E, Antonini M, Grillo G, Tabasso S, Quagliotto P, et al. Product yield, quality and energy in the hydrolysis of urban bio-waste compost from laboratory-scale runs. Journal of Cleaner Production. 2018;**170**:1484-1492. DOI: 10.1016/j.jclepro.2017.09.196

[38] Azwar E, Wan Mahari WA, Chuah JH, Vo DVN, Ma NL, Lam WH, et al. Transformation of biomass into carbon nanofiber for supercapacitor application—A review. International Journal of Hydrogen Energy. 2018;**43**:20811-20821. DOI: 10.1016/j.ijhydene.2018.09.111

[39] Kanoksilapatham W, Intagun W, Review A. Biodegradation and applications of keratin degrading microorganisms and keratinolytic enzymes, focusing on thermophiles and thermostable serine proteases. American Journal of Applied Sciences. 2017;**14**:1016-1023. DOI: 10.3844/ajassp.2017.1016.1023

[40] Belarmino DD, Ladchumananandasivam R, Belarmino LD, de Pimentel JRM, da Rocha BG, Galvão AO, et al. Physical and morphological structure of chicken feathers (keratin biofiber) in natural, chemically and thermally modified forms. Materials Sciences and Applications. 2012;**03**:887-893. DOI: 10.4236/msa.2012.312129

[41] Chaitanya Reddy C, Khilji IA, Gupta A, Bhuyar P, Mahmood S, Saeed KA, et al. Valorization of keratin waste biomass and its potential

applications. Journal of Water Process Engineering. 2021;**40**:101707. DOI: 10.1016/j.jwpe.2020.101707

[42] Gopinath V, Saravanan S, Al-Maleki AR, Ramesh M, Vadivelu J. A review of natural polysaccharides for drug delivery applications: Special focus on cellulose, starch and glycogen. Biomedicine & Pharmacotherapy. 2018;**107**:96-108. DOI: 10.1016/j.biopha.2018.07.136

[43] Anup N, Chavan T, Chavan S, Polaka S, Kalyane D, Abed SN, et al. Reinforced electrospun nanofiber composites for drug delivery applications. Journal of Biomedical Materials Research Part A. 2021;**109**:2036-2064. DOI: 10.1002/jbm.a.37187

[44] Yadav M, Goel G, Hatton FL, Bhagat M, Mehta SK, Mishra RK, et al. A review on biomass-derived materials and their applications as corrosion inhibitors, catalysts, food and drug delivery agents. Current Research in Green and Sustainable Chemistry. 2021;**4**:100153. DOI: 10.1016/j.crgsc.2021.100153

[45] Zhang Z, Ji D, He H, Ramakrishna S. Electrospun ultrafine fibers for advanced face masks. Materials Science & Engineering R: Reports. 2021;**143**:100594. DOI: 10.1016/j.mser.2020.100594

[46] Ilyas RA, Sapuan SM, Ibrahim R, Atikah MSN, Atiqah A, Ansari MNM, et al. Production, processes and modification of nanocrystalline cellulose from agro-waste: A review. Nanocrystalline Materials. 2020;**1**:89-95. DOI: 10.5772/intechopen.87001

[47] Barhate RS, Ramakrishna S. Nanofibrous filtering media: Filtration problems and solutions from tiny materials. Journal of Membrane Science. 2007;**296**:1-8. DOI: 10.1016/j.memsci.2007.03.038

[48] Asadi H, Ghaee A, Nourmohammadi J, Mashak A. Electrospun zein/graphene oxide nanosheet composite nanofibers with controlled drug release as antibacterial wound dressing. International Journal of Polymeric Materials and Polymeric Biomaterials. 2020;**69**:173-185. DOI: 10.1080/00914037.2018.1552861

[49] Adam AA, Dennis JO, Al-Hadeethi Y, Mkawi EM, Abdulkadir BA, Usman F, et al. State of the art and new directions on electrospun lignin/cellulose nanofibers for supercapacitor application: A systematic literature review. Polymers (Basel). 2020;**12**:1-36. DOI: 10.3390/polym12122884

[50] Wen Y, Kok MDR, Tafoya JPV, Sobrido ABJ, Bell E, Gostick JT, et al. Electrospinning as a route to advanced carbon fibre materials for selected low-temperature electrochemical devices: A review. Journal of Energy Chemistry. 2021;**59**:492-529. DOI: 10.1016/j.jechem.2020.11.014

[51] Svinterikos E, Zuburtikudis I, Al-Marzouqi M. Electrospun lignin-derived carbon micro- and nanofibers: A review on precursors, properties, and applications. ACS Sustainable Chemistry & Engineering. 2020;**8**:13868-13893. DOI: 10.1021/acssuschemeng.0c03246

[52] Pérez-Madrigal MM, Edo MG, Alemán C. Powering the future: Application of cellulose-based materials for supercapacitors. Green Chemistry. 2016;**18**:5930-5956. DOI: 10.1039/c6gc02086k

[53] Zhai Y, Dou Y, Zhao D, Fulvio PF, Mayes RT, Dai S. Carbon materials for chemical capacitive energy storage.

Advanced Materials. 2011;**23**:4828-4850. DOI: 10.1002/adma.201100984

[54] Guan L, Yu L, Chen GZ. Capacitive and non-capacitive faradaic charge storage. Electrochimica Acta. 2016;**206**:464-478. DOI: 10.1016/j.electacta.2016.01.213

Section 2

Electrospun Functional Materials for Applications in the Food Industry, Electronics and Biomedical Engineering

Chapter 5

Electrospinning of Fiber Matrices from Polyhydroxybutyrate for the Controlled Release Drug Delivery Systems

Anatoly A. Olkhov, Svetlana G. Karpova, Anna V. Bychkova, Alexandre A. Vetcher and Alexey L. Iordanskii

Abstract

The submission provides an overview of current state of the problem and authors' experimental data on manufacturing nonwoven fibrous matrices for the controlled release drug delivery systems (CRDDS). The choice of ultrathin fibers as effective carriers is determined by their characteristics and functional behavior, for example, such as a high specific surface area, anisotropy of some physicochemical characteristics, spatial limitations of segmental mobility that are inherent in nanosized objects, controlled biodegradation, and controlled diffusion transport. The structural-dynamic approach to the study of the morphology and diffusion properties of biopolymer fibers based on polyhydroxybutyrate (PHB) is considered from several angles. In the submission, the electrospinning (ES) application to reach specific characteristics of materials for controlled release drug delivery is discussed.

Keywords: polymer therapeutic systems, electrospinning, biopolymer fibers, nonwoven fibrous matrices, polyhydroxybutyrate, controlled diffusion transport, controlled release, morphology, segmental mobility

1. Introduction

The modern paradigm of targeted drug delivery to the organ of interest, its part, or the target cell is based on the use of biocompatible carriers, the sizes of which are in the submicron (nanometer) range. Today, functional carriers of biologically active compounds and absorbents with a high specific surface area are widely used in biomedicine in the form of bioceramic minerals, polymer therapeutic systems, framework structures for cell engineering, nanoscale hybrid means of targeted drug transport, and a number of innovative systems that simultaneously function as implants and carriers of biologically active compounds [1–5]. Among these functional materials, the greatest practical and commercial advantages, in the near future, will be obtained by hybrid micro- and nanoparticles [6, 7], as well as composite ultrathin fibers [8, 9].

Micro- and nanoparticles based on biodegradable materials offer a wide range of applications as components of innovative forms or independent systems for drug delivery in the implementation of anti-inflammatory and antitumor therapy, wound healing, and thrombolytic therapy. The use of magnetic nanoparticles as part of innovative systems can facilitate targeted drug delivery, reducing the required dosage of a drug, and provide a prolonging effect of the activity of biological macromolecules, for example, enzymes, and the ability to visualize drug delivery processes [10–12].

The employment of biodegradable and biocompatible particles and fibers creates additional advantages in the development of a new generation of implanted therapeutic systems. Therefore, nowadays, the main part of the work aimed to develop biodegradable compositions in medicine is devoted to the creation of materials for tissue engineering. Modern metal implants (titanium and its alloys, stainless steel, etc.) are characterized by rigidity and endurance and do not cause an immune response but do not have the properties of osteoinduction and osteoconductivity. Often their application turns to add an additional surgical intervention to remove the implant. This increases additional risks and the duration of the patient's recovery process, and the therapy aimed at restoring the bone tissue and its environment. Therefore, inorganic materials analogous to bone tissue, such as apatite and complex composite biodegradable forms that provide the transport of biological molecules and synthetic drugs, are objects of fundamental and applied research.

One of the innovative ways to obtain ultrathin fibers and fibrillar nanomaterials is the method of electrospinning (ES) of polymer solutions and melts [13, 14]. The development of technology for the formation of nanofibers in an electrostatic field makes it possible to create materials of various shapes and morphology with a high specific surface area and porosity, with adequate mechanical properties, and a wide range of structural, dynamic, and diffusion characteristics. Fibrillar matrices and mats formed by the obtained nanofibers create favorable conditions for free migration and proliferation of cells in the three-dimensional space of frame structures, and, accordingly, provide a high integration affinity of the material for living tissues of the body [15]. They are actively used in the design of biosensors, nanofilters, for wound therapy, to immobilize enzymes, in the creation of prolonged and targeted drug delivery systems, and other areas of modern biology and medicine [16–20].

It should be noted that in contrast to a great availability published data on the nanoparticles applications, the data on biodegradable and bioresorbable nanofibes is rather rare [21–23]. About a third of publications analyze the peculiarities of the ES technology, and the rest considers their characteristics, functional behavior *in vitro* and *in vivo*, as well as preclinical and clinical trials.

2. Fields of application of biopolymer forms for therapeutic purposes

Biodegradable compositions are also applied in solving the problem of wound healing, which may include a set of subtasks—stopping bleeding, preventing inflammation, proliferation, and tissue reconstruction. Chronic wounds are characterized by high protease activity, infection, inflammation, and hypoxia [24–27]. A wide range of wound healing materials is currently being developed. These are unfilled materials based on siloxysiloxane, dextran, urethane, collagen, etc.; materials including stem cells or vitamin E based on hyaluronic acid; materials based on extracellular matrix proteins (fibrin, fibronectin, collagen, etc.) for tissue reconstruction and angiogenesis; and materials including antibiotics and other drugs [28]. Materials in the form of films, hydrogels, foams, multilayer compositions, etc., have already received clinical use. Chitosan, known for its hemostatic and antibacterial activity, is used in some currently developed composite forms for accelerated wound healing and drug delivery for matrix formation and tissue reconstruction. In [29], chemically stable composite materials for wound coverage were based on two polysaccharides—cellulose and chitosan combining mechanical strength (cellulose) and the ability to stop bleeding, cleanse and heal wounds, deliver drugs, and overall bactericidal properties (chitosan).

It is known that the regeneration of cartilage tissue is directly related to the number of chondrocytes per unit of its mass, and the introduction of a suspension of chondrocytes into the damaged areas is currently considered a promising therapeutic approach. Foreign chondrocytes, being introduced into the joint cavity, do not cause a rejection reaction, because they have limited immunogenic activity. Much attention is now being paid to the development of composite hydrogels for injections—research is being carried out using photopolymerization, chemical crosslinking of molecules in the composition of the composite with carbodiimide, glutaraldehyde, genipin, and adipic dihydrazide. In the first case, the use of a photosensitizer and radiation is required, which limits the applicability of the approach, and in the case of crosslinkers, the main problem is their toxicity toward cells. In [30], a new representative of composite hydrogels was obtained based on chitosan and oxidized hyaluronic acid. The formation of the gel proceeded through the formation of a Schiff base between the amino and aldehyde groups of polysaccharide derivatives, N-succinyl-chitosan (S-CS) and hyaluronic acid aldehyde, and did not involve the use of a chemical crosslinking agent. The potential of using such composite hydrogels as scaffold structures for injections has been shown using the example of cartilage tissue cells in joints. In this study, the high content of chitosan led to a decrease in the rate of degradation of the composite.

Let us consider several examples of the use of composite particles based on biodegradable polymers. In [31], hydrogels based on chitosan and gelatin are used for long-term administration of the antiglaucoma drug timolol maleate and can reduce the side effects of its use. Hydrogels based on sulfonated chitosan and heparin-like chitosan (containing carboxymethyl and sulfate groups) increase blood clotting time [32], prevent protein and platelet adsorption on the membrane intended for blood purification [33], and prevent complement activation [34]. Hydrogels consisting of chitosan, heparin, and poly (γ-glutamic acid) with different ratios of components and loaded with superoxide dismutase were used to create a wound dressing with antioxidant properties [35], which had the potential to treat chronic trauma in diabetes and proved to be a promising wound healing agent. Heparin-loaded hydrogel based on photosensitive hydroxyethylchitosan promoted a long-term effect of lowering intraocular pressure after surgery to eliminate glaucoma [36]. While the possibilities of transporting the anticoagulant heparin as a constituent element of particles are being actively studied, a much smaller number of works are devoted to the transport of proteins associated with the functioning of fibrinolytic and anticoagulant systems—tissue plasminogen activator (tPA), streptokinase, and urokinase. The formation of complexes between enzymes and biodegradable polymers proved to be an effective way to overcome the short biological half-life of enzymes in the bloodstream, and the inclusion of magnetic nuclei in the particles revealed the potential of using biodegradable materials for targeted drug transport and diagnostics. Magnetic nuclei delivered on their envelope (for example, the composition of polyethylene glycol (PEG) and chitosan [37]) tPA or streptokinase promote thrombolysis controlled by an external magnetic field. Hydrogels based on chitosan and alginate with incorporated magnetic particles were investigated as carriers of the anticancer drug matrine for oral administration [38, 39]. The obtained systems provide pH-sensitive release and targeted delivery under the influence of weak magnetic fields. The inclusion of magnetic particles in systems based on hyaluronic acid allows to detect the processes of tissue regeneration and biological delivery of drugs with the participation of constructed systems [40].

The areas of possible application of biodegradable forms are not limited to the aforementioned examples. New designs of cardio stents with biodegradable or inert coatings are capable of providing targeted transport of such cardiac drugs as paclitaxel, sirolimus, tacrolimus, etc. These drugs are responsible for suppressing cell proliferation; therefore, their prolonged delivery significantly reduces the number of restenosis incidents as compared to the first generation of spring metal stents. Moreover, a nano-level modified stent surface for controlled cell adhesion is desirable. In urology, a separate group of polymer implants consists of representatives of natural biodegradable polymers—poly-α-hydroxy acids (polylactides and their copolymers) and poly-β-hydroxyalkanoates (polyhydroxybutyrate (PHB), and its copolymers with oxyvalerate, oxyhexanoate, etc.), and the base for neurological implants are biodegradable polymers such as polylactides or polyhydroxyalkanoates [41]. For urological applications, a prolonged release of bactericidal drugs is required, whereas for nerve tissue regeneration, a drug activates the development of a nerve impulse.

Modern works also include the creation of systems with a reverse response, an example of which is the form that releases insulin depending on the concentration of glucose in the blood [42]. The possibility of visualizing the area of development of the pathological process and monitoring the course of the treatment process is presented with the use of isotopes, fluorescent dyes, quantum dots, magnetic nanoparticles, and other markers [43]. Biodegradable innovative materials, thus, have received a wide range of possible applications in the treatment of various diseases (**Figure 1**) in the form of various medical forms—macroobjects, films, micro-, and nanoparticles, etc. A wide variety of required compositions and morphology of matrices of medical forms contributes to the fundamental and applied issues associated with the use of natural

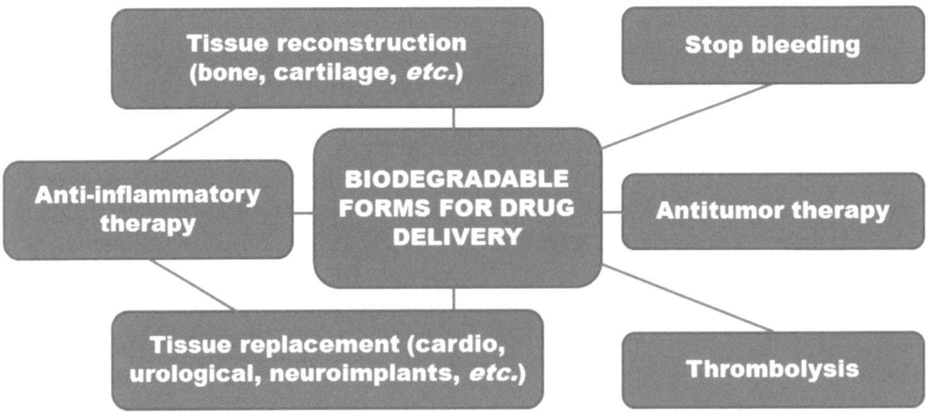

Figure 1.
The main areas of pharmaceutical forms' application for the delivery of low- and high-molecular-weight pharmaceuticals based on biodegradable macromolecules.

and synthetic materials for the delivery of low- and high-molecular-weight biologically active substances (BASs). More details on the possible chemical composition of matrices for drug transport and methods of drug inclusion in the matrix will be presented in the next section.

3. The main components of biodegradable innovative forms of drug delivery

Let us consider the main components of biodegradable materials that make up innovative forms using the example of collagen, hyaluronic acid, carboxymethyl cellulose (CMC), chitosan, and other natural and synthetic macromolecules, the structural formulas of which are shown in **Figure 2**, most often used for bone tissue regeneration [43].

Chitosan is a polysaccharide of D-glucosamine and N-acetyl-D-glucosamine, linked by β (1 → 4) -glycosidic bonds. Chitosan is obtained by cleavage of the acetyl group from chitin. The deacetylation reaction can also be accompanied by the rupture of the glycosidic bonds of the polymer. The degree of deacetylation of chitosan and its molecular weight predetermines its reactivity and properties, causing the structural heterogeneity of macromolecules. The solubility of chitosan depends on the pH (it is insoluble at neutral values), and its salts, for example, chitosan hydrochloride and chitosan glutamate, are soluble at any pH.

The hydrophilicity of chitosan and its positive charge facilitate its reactions with negatively charged macromolecules and polyanions, which are both components of the biodegradable composite matrix and functional molecules.

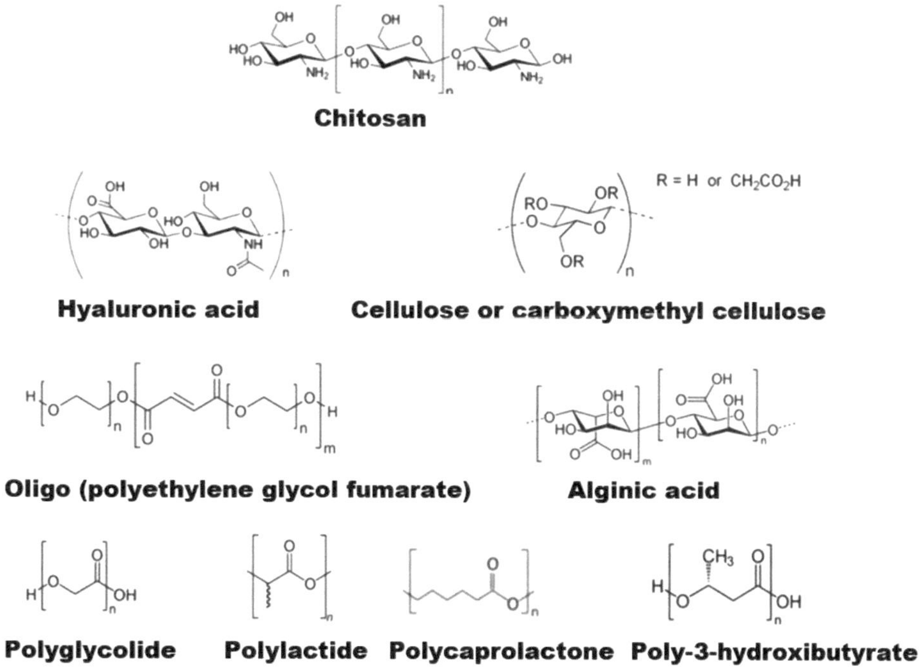

Figure 2.
Structural formulas of some biodegradable macromolecules used for targeted transport of low- and high-molecular pharmaceuticals.

This makes it possible to use sol-gel processes for drug binding. The positive charge of chitosan also promotes adhesion to the mucous membranes of the body, which makes it possible to use it for the transport of drugs through the mucous membrane. Thus, the features of chitosan are pH-dependent solubility (at pH <5 and basic pH) and insignificant toxicity due to —NH_2 groups (it undergoes enzymatic degradation in vivo, and its degradation products are involved in the metabolic cycle). CMC can be used together with cellulose and is similar in structure to chitosan. The sodium salt of CMC is a water-soluble polymer. Cellulose is a polysaccharide with the formula $(C_6H_{10}O_5)_n$, in which the D-glucose residues are linked by β (1 → 4)-glycosidic bonds. Cellulose undergoes dissolution under the action of ionic liquids (organic salts, liquid at room temperature). In [29], such a solvent was used to obtain a composite material for wound dressings based on two polysaccharides – cellulose and chitosan.

Fibrinogen and fibrin are the next examples of protein molecules that can be components of micro- and nanoscale scaffolds. These proteins are present in the blood and are the main participants in the process of coagulation, which determines their high ability to interact with damaged tissues and cells. To solve problems associated with tissue regeneration, proteins can be modified or introduced into more complex systems, including, for example, growth factors or other proteins, as well as stem cells [44]. Fibrin gel formation can be initiated directly at the site of injury using particles on the surface of which thrombin is fixed [45].

The main protein of silkworm silk, fibroin, and the skeleton silk of the spider web, spidroin, have crystalline parts responsible for high tensile strength and amorphous parts that provide elasticity, which makes them versatile materials for use in tissue engineering, pharmacy, and medicine, regardless of the type of construction. The breakdown products of silk fibroin and spidroin are amino acids, which act as an additional building material for tissue regeneration [46].

Hyaluronic acid is a hydrophilic, non-immunogenic, biodegradable glucosaminoglycan—a polymer consisting of D-glucuronic acid and D-N-acetylglucosamine residues, linked alternately by β-1,4- and β-1,3-glycosidic bonds, which, in combination with other osteoconductive molecules, promotes bone growth. It has a high content in extracellular matrices, is a component of articular cartilage, and is part of the skin. Amphiphilic derivatives of hyaluronic acid promote its self-assembly in a core-shell nanogel for the transport of hydrophobic pharmaceuticals, e.g. anticancer.

For the transport of low- and high-molecular-weight therapeutic substances, "depot carriers" are used based on sodium alginate and its compositions with pectin, sodium hyaluronate, etc. Salts of alginic acid (alginates), in contrast to the polysaccharide acid itself, form colloidal solutions in water and have antimicrobial and hemostatic action.

Unlike natural macromolecules, synthetic biodegradable molecules such as poly-α-hydroxy acids such as hydrophobic polylactides (PLAs) and more hydrophilic polyglycolides (PGAs), as well as poly-β-hydroxyalkanoates (PHBs and its derivatives) do not elicit a biological response and are widely distributed as components of innovative forms for drug delivery. Degradation of polymers proceeds through hydrolysis of ether groups, and the rate and products of degradation are determined by the composition, structure, and molecular weight of molecules, as well as the content of ions and enzymes in biological fluids. Polymeric materials of the class of poly-β-hydroxyalkanoates or poly-α-hydroxy acids degrade to nontoxic products—CO_2 and H_2O [41]. Also commonly used are aliphatic polyesters, that is, polycaprolactones (PCLs). Blends of polymers or copolymers are often the basis for materials intended for the restoration of bone and cartilage tissue. The most popular copolymer is polylactide-co-glycolide (PLGA), the ratio of the components of which affects the hydrophilicity and biodegradability characteristics. During copolymerization, the degree of crystallinity decreases, as a result of which hydrolytic destruction proceeds faster [47]. Other common synthetic macromolecules for bone tissue regeneration are polypropylene fumarate (PPF), polyanhydrides, and polyphosphazenes [26]. Oligo (polyethylene glycol fumarate) (OPF) based on PEG is able to biodegrade, while PEG is not [48]. Degradation of PPF and OPF is due to the fumaric acid residue in the polymers. Copolymerization of macromolecules, for example, polyanhydrides, can be carried out to increase the hydrophobicity of materials and reduce the rate of their biodegradation. The combination of natural and synthetic materials in the compositions provides a wide range of properties of the systems and the possibilities of their use. An example of a natural material used for the transport of various drugs in the composition of composite forms is gelatin [49].

In innovative materials, the simultaneous use of natural and synthetic macromolecules creates unique opportunities for the delivery of various pharmaceuticals and ensuring the required release profiles. As in the case of synthetic ones, when using natural hydrogels, chemical and physical crosslinking of macromolecules allow one to regulate the diffusion of therapeutic substances included in biodegradable forms, and the biodegradation of macromolecules is also determined by the degree of their crosslinking and the type of crosslinking agent. In this case, the pharmaceuticals themselves can be covalently and non-covalently associated with biodegradable compositions, in particular, the inclusion of pharmaceuticals in the composition of $Ca_3(PO_4)_2$ particles is carried out physically or chemically [24], and the release of pharmaceuticals can be controlled by the chemistry of $Ca_3(PO_4)_2$, the porosity of the material, the surface area of the hydrogel particles, their charge surface, and crystallinity. In the case of covalent immobilization of macromolecular pharmaceuticals (for example, proteins), the release mechanism includes chemical/enzymatic cleavage of the active substance. The introduction of an antibiotic in the treatment of osteomyelitis can be provided by direct mixing of the drug powder with a bone graft or soaking the bone graft in an antibiotic solution [26]. Similarly, for the prolonged delivery of streptokinase, systems obtained by mixing the protein with chitosan were used [50]. Urokinase and streptokinase showed their activity as thrombolytics both on the surface and in the bulk of particles from chitosan and tripolyphosphate, providing delivery, and the indicators in both cases were higher than in the protein solution [51]. The introduction of pharmaceuticals into composite structures from a solution after their creation is one of the approaches that damage pharmaceuticals to the least extent. Chondrocytes in [30] were introduced into the hydrogel from a solution in which they were resuspended. Another example of a technique that provides a minimal effect on the drug structure is the approach using the state of a supercritical fluid, which provides a one-stage filling of porous matrices with medicinal substances. Approaches are known that include the formation of a special drug complex, for example, with cyclodextrins, for its use in a sol-gel process, during which biodegradable composite structures with pharmaceuticals incorporated into them are created [52]. Pharmaceuticals can also be adsorbed on the surface of implanted structures and coated with biodegradable polymeric materials to ensure their long-term release, which is realized, for example, in [25] during transfection. A hydrophobic drug can also be encapsulated in an oil core covered with a biodegradable shell, including, for example, chitosan [53]. Drug release can be divided by type into diffusion-controlled, controlled by chemical processes or matrix swelling, and controlled by external processes or devices.

Most of the listed approaches for the inclusion of pharmaceuticals in various matrices can also be implemented when nanoparticles are included in hydrogels. The particles are introduced into the compositions after the formation of the gel and at the stage of gel formation, with or without the use of covalent binding processes. Nanoparticles in the composition of hydrogels provide a change in their mechanical properties and swelling characteristics and are also able to impart magnetic, optical properties, electrical conductivity, and improved antimicrobial properties to systems [54]. Hybrid hydrogels may include, for example, carbon-based nanoparticles, inorganic particles, and nanoparticles of semiconductors, nanoparticles of metals and their oxides, polymers, and liposomes. For example, the introduction of magnetite nanoparticles into the hydrogel is provided due to their stabilization by oleic acid and the amphiphilic nature of the main component of the hydrogel-modified hyaluronic acid [40]. Pyrene ligands contribute to the formation of hydrogel particles with a "core/shell" structure. In this case, magnetic particles bound to thrombin were employed to form fibrin-based skeletal structures [45]. Solid colloidal nanoparticles can be the core for a polyelectrolyte shell deposited using layer-by-layer adsorption technologies [40], as well as be part of the shell itself, providing magnetically controllable particles. For example, the authors [55] modified hollow microcapsules obtained from dextran sulfate and poly-L-lysine with maghemite particles.

Thus, a large group of biological and synthetic macromolecules is used to create innovative forms for the drug delivery. Although a more detailed overview of the methods of drug encapsulation and biodegradable matrices composition cannot be presented within the required limits of this chapter, it is obvious that the chemical composition of materials combined with the features of medical forms created on their basis (which include characteristics such as size, porosity, the presence of covalent crosslinking, and stimulus sensitivity) and the method of drug encapsulation predetermine the different fate of biodegradable materials in biological media and different profiles of the release.

4. Structure and morphology of ultrathin fibers as drug carriers

4.1 Influence of dipyridamole (DPD) on the structure and segmental mobility of a biopolymer in ultrathin fibers

Among the biopolymers used in dentistry, traumatology, orthopedics, cellular engineering, surgery, along with polylactides, the most common is, probably, PHB, the main representative of the biopolymers of the polyhydroxyalkanoate family [56]. This biopolymer is a product of microorganisms' biosynthesis. It has high biocompatibility, the ability to rapid biosorption without the formation of toxic products, and increased resistance to oxidative degradation [57, 58].

In many previous reports, the structure formation of fibrillar materials based on PHB-containing dipyridamole (DPD) [59], chitosan [60], titanium dioxide and silicon nanoparticles [61], iron (III)-chlorporphyrin complexes [62], etc. was considered. In these works, the influence of low-molecular-weight substances on the structure of the crystalline and amorphous phases of PHB fibers was shown. This section will present a review of the structure and properties of ultrathin PHB fibers obtained by ES, containing immobilized drug – dipyridamole (DPD).

The structure of intercrystalline regions in fibers containing drug is directly related to the state of the crystalline phase of the carrier polymer by the spatial organization of the pass-through chains [63]. The latter, in the case of a high degree of crystallinity, experiences deformation and conformational spatial difficulties. Therefore, it is necessary to distinguish two alternative situations, the first: when the degree of crystallinity is low and the distance between the crystallites and polymer lamellae is large, so that a relatively low concentration of drug is evenly distributed between polymer crystals and should have little effect on the conformation of uncrystallized polymer molecules. An alternative situation arises when conditions are created in highly crystalline polymers (such as PGB, PLA, and polyamide-6) to realize the maximum degree of crystallinity, for example, as a result of solvent plasticization or as a result of temperature annealing. In this case, the proportion of polymer segments included in noncrystalline regions is relatively small, the mechanical and diffusion behavior of the polymer is determined by the state of the elongated flow chains, and the concentration of the introduced drug, related to the volume of the intercrystalline phase, may exceed its thermodynamic solubility. Then the excess of the low-molecular-weight component is displaced from the polymer volume with the formation of an independent drug phase on the fiber surface, and the drug molecules remaining in the volume are potentially capable of influencing the conformation and dynamics of polymer molecules in the intercrystalline regions of the polymer.

The structure and molecular dynamics of these regions in biopolymer fibers can be effectively investigated by the electron paramagnetic resonance (EPR) method using the microprobe technique of stable nitroxyl radicals developed at the IHF RAS (Moscow, Russian Federation) [64]. The satisfactory agreement between the calculated and experimental results demonstrated earlier indicates, first of all, the effectiveness of the selected two-phase model of the intercrystalline space of the PHB fiber, which consists of more and less dense regions with corresponding different rotational mobilities of a stable radical in these regions.

An increase in the concentration of dense regions in the presence of a drug was observed for the composite system PHB-DPD. The corresponding calculations revealeded that the volume fraction of more dense regions in the intercrystalline space of PHB fibers is much higher than the content of less dense regions. Moreover, when pharmaceuticals are added to the fiber, it continues to grow insignificantly following the sequence 0.90 (0) $<$ 0.93 (1) $<$ 0.94 (3) $<$ 0.95 (5), where the numbers in parentheses indicate the percentage of DPD mass concentration. This result seems quite natural if we take into account that the specific enthalpy of melting of PHB changes in the same sequence, reflecting, as in the previous case, the degree of its crystallinity.

As the concentration of the dense fraction in the intercrystalline regions of PHB increases, a corresponding decrease in the rotation rate of the radical is observed, and, therefore, the segmental mobility of macromolecules slows down. For the slow component of the rotational mobility, similar changes in the values of the correlation time in the fiber are observed, namely, the value of this dynamic parameter increases with an increase in the DPD content in the sequence 6.6×10^{-9} s (0) > 7.1×10^{-9} s (1) > 8.8×10^{-9} s (3) > 9.0×10^{-9} s (5%), which indicates a slowdown in the molecular mobility of the radical and, accordingly, a decrease in the molecular mobility of PHB chains in intercrystalline areas. The correlation time for the fast component in all samples, except PHB with 5% DPD (7×10^{-10} s), was 2.4×10^{-10} s, i.e. more than an order of magnitude lower than the previous values. Consequently, a change in the crystalline phase of the polymer affects the dynamics of chains in denser regions of the intercrystalline space and practically does not affect the fast component of their mobility in less-dense intercrystalline regions. These results can be explained within the framework of the model of the heterogeneous (binary) structure of intercrystalline regions of polymers with a high degree of crystallinity [65, 66]. The perfection of the crystal structure of nanofibers depends on the conditions of electrospinning, namely, on the rate of solvent desorption from the formed fiber and the temperature regime of its cooling [63]. Indeed, fragments of the polymer chain with a predominance of the straightened conformation are mainly involved in the formation of paracrystalline regions and the recrystallization of the polymer. Therefore, probe molecules with a correlation time τ in paracrystalline regions are sensitive to changes in the degree of crystallinity of PHB. On the contrary, the molecules of the same probe, located at a considerable distance from the crystals in regions with fast segmental mobility and approaching in their dynamic characteristics to the amorphous phase of the polymer, are practically unaffected by the crystalline phase of the fiber and, within some limits, retain their constant value.

4.2 Influence of thermal heating of PHB fibers containing a dipyridamole on the molecular dynamics of polymer chains

During ES, as a result of difficult cooling and curing conditions of ultrathin fibers, their polymer structure can be far enough from the state of thermodynamic equilibrium. The imperfection of the crystalline phase and morphology is manifested in the insufficient orientation of the segments in the fiber, as well as in an atypically low degree of crystallinity [63]. Thermal annealing several tens of degrees below the melting point of PHB (annealing temperature 140°C) allows to sharply intensify segmental mobility and transfer the system to a more thermodynamically equilibrium state. Indeed, the time of thermal annealing at 2 h for PHB combinations with pharmaceuticals in the absence or low drug content for a number of samples significantly affects the rotation dynamics of the TEMPO radical. Whereas for highly crystalline fibers (with 3 and 5% DPD), this process has little effect on the rotational mobility of the probe, reflecting the segmental mobility of the chains (**Figure 3**).

One of the possible mechanisms of PHB recrystallization upon thermal annealing of the fiber is that a part of the chain segments with a high degree of orientation β,

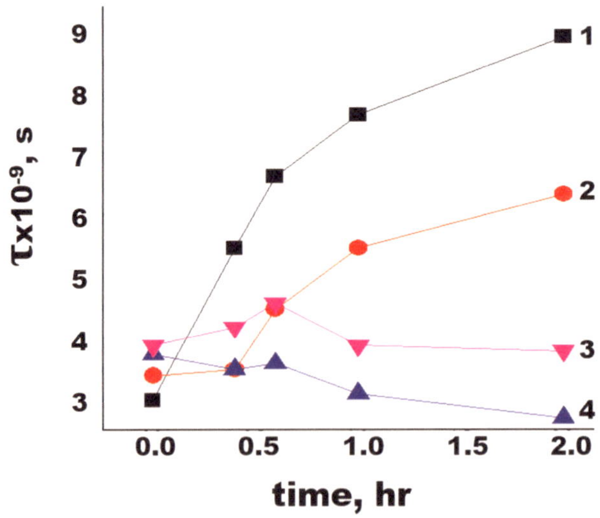

Figure 3.
Dependences of the effective correlation time (τ) on the annealing time at 140°C: 1—PHB, 2—PHB with 1%, 3—PHB with 3%, 4—with 5% DPD.

which was acquired during the electrospinning process, reaches values greater than the critical value $h/L \sim \sqrt{2/l}$, where h is the distance between chain ends, L is the persistent chain length, and l is the length of a single segment [67]. As a result, the sorption capacity of the fiber due to an increase in the degree of crystallinity decreases. Indeed, measurements of the concentration of a stable radical in the fibers also show a sharp decrease in this value after annealing for 2 h. For example, in the initial PHB fibers, the concentration of the radical after absorption from vapors was 8×10^{15} spin/g, while this value after annealing for 2 h decreased to 1.4×10^{15} spin/g. A similar picture was observed for other biopolymer fibers containing a porphyrin complex.

The correspondence is observed between the content of the crystalline phase of PHB (structural characteristic) and the value of the correlation time (dynamic characteristic), so that molecular mobility decreases as the total volume of intercrystalline regions decreases due to an increase in crystallinity of PHB and, consequently, the involvement of an increasing number of mobile polymer segments in dense polymer regions with low mobility and high values of τ. Obtained results allow to conclude that, in the presence of a drug, from a PHB solution by ES, ultrathin fibers of various geometries with structures of various degrees of equilibrium and perfection are formed, which is true both for the crystalline phase and for intercrystalline regions, where redistribution between the amorphous and paracrystalline states of PHB is possible. The nonequilibrium of intercrystalline regions for PHB with the absence or low content of biologically active substances (DPD) is confirmed by a change in the rotational mobility of a stable radical and an increase in crystallinity. All results, including the effect of drug concentration on the fiber shape and its dynamic characteristics, are in good agreement with the thermophysical parameters of the system and should be directly applied if describing the directed and prolonged transport of bioactive compounds.

A natural continuation of these studies is the transition from the structural characteristics of ultrathin fibers to the study of the diffusion kinetics of a drug in their bulk. In the next section, the results of the dependence of diffusion transport as the main process responsible for controlled drug release on the geometric dimensions of the fiber, its crystallinity and the porosity of fibrillar films will be presented. The obtained results will be used to consider diffusion kinetics, which, in combination with the analysis of segmental dynamics, represent two fundamental processes that determine the rate and mechanism of controlled drug release from fibrillar therapeutic systems.

5. Diffusion and controlled drug release in ultrathin PHB fibers and fibrillar films based on them

The process of controlled targeted delivery of biologically active substances, for example, anti-inflammatory and antidiabetic pharmaceuticals or growth hormone [68–70], cannot be adequately described without considering their diffusion. In this regard, despite the impressive technological advances in the creation of ultrathin polymer fibers for various applications, the solution of diffusion and enzymatic-hydrolytic problems in the systems monofilament—biologically active substance (BAS) and fibrillar matrix—BAS is found in an extremely limited number of works (e.g. [71–73]) and requires in-depth analysis both experimental and theoretical levels. Consideration of the earlier-mentioned diffusion-kinetic problem was carried out using the example of ultrathin fibers with an encapsulated drug. Combining the results of transport in fibers and fibrillar membranes with the structural and dynamic characteristics of a biopolymer carrier, a consistent controlled release model is proposed that satisfactorily describes the combination of BAS diffusion and hydrolysis kinetics in an innovative therapeutic system based on a typical biodegradable polyester (PHB).

In **Figure 4**, typical kinetics of the release of dipyridamole (DPD) from fibers of different geometry consisting of ellipse-like and cylindrical structures and with different drug concentrations is plotted. As in the case of the "monolithic PHB film—drug" system [74], for the fibrillar matrix formed by ultrathin PHB fibers, the kinetic release profiles have two characteristic kinetic regions of linear and nonlinear form. The bimodal nature of controlled release is especially pronounced for fibers with a higher DPD content in the fiber (3 and 5 wt%). Under our proposed model, the initial nonlinear section mainly reflects a diffusion process with a characteristic drug diffusion coefficient, while the linear region corresponds to a kinetic process reflecting fiber degradation with partial loss of its mass. Briefly, the essence of this process is determined by the onset of hydrolysis of the ester functional groups of PHB. In the process of hydrolytic destruction, the drug encapsulated in the fiber passes into the surrounding aqueous solution not only as a result of diffusion but also as a result of partial fiber disintegration by the mechanism of surface destruction or erosion [75].

In kinetic measurements, the fiber sample was immersed in a 60% aqueous solution of ethanol and the optical density of DPD samples was determined with a periodic sampling. The interval of sampling depended on the fiber composition and, accordingly, on the rate of drug release; it was from 1 to 30 min. The experiments lasted from several hours to several days. Termination of an increase in the optical density of DPD exhibiting two characteristic peaks at λ = 410 nm and a more intense peak at λ = 292 nm with an extinction coefficient of 31,260 L/(mol $\times$ cm) indicated that the drug release was completed.

With this formulation of the problem, the kinetic profile of drug release is described by the following system of equations. During the time interval Δt, the cumulative mass of the drug released from the fibrillar film by the time t ($\Delta M_d(t)$) is

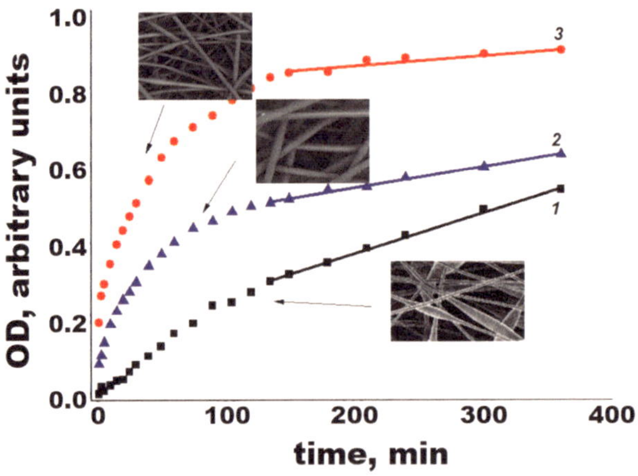

Figure 4.
Typical kinetic profiles of controlled release of DPD from PHB fibers. The concentration of DPD is 1 (1), 3 (2), and 5 wt% (3). scanning electron microscopy (SEM) photomicrographs illustrate the shape of the fibers; magnification 1000×.

the sum of two terms: the amount of drug that entered the solution volume by the diffusion mechanism (ΔM_D) and the amount of the drug that passed into solution as a result of partial loss of fiber mass ΔM_f, which contained ΔM_k grams of drug immobilized by polymer molecules and incapable of diffusion in a polymer medium. The above reflects a simple balance of the change in the mass of the active component in the fiber, which reflects the Eq. (1);

$$\Delta M_d/\Delta t = \Delta M_D/\Delta t + \Delta M_k/\Delta t \quad (1)$$

With a constant volume of the surrounding solution V and intensive stirring, external diffusion restrictions can be neglected. In this case, for a fibrillary film, by analogy with a monolithic nonporous film [75], one can write a differential equation showing the simultaneous contribution of drug diffusion and PHB hydrolysis by the zero-pillage mechanism:

$$(\partial M_d/\partial t)/V = \partial C_d/\partial t = D_{eff}\left[\partial^2 G_d/\partial x^2\right] + k_h \quad (2)$$

where C_d and G_d are the total concentration of the drug that entered the external volume V by the time t and the concentration of the mobile drug fraction capable of diffusing in the polymer sample with the corresponding effective diffusion coefficient D_{eff}, independent of the coordinate and time, and kh is the fiber mass loss constant, mainly due to the hydrolysis of the ester groups of PHB.

In a fibrillar film formed by a random interlacing of ultrathin fibers, D_{eff} is determined by two coupled processes: the diffusion mobility of drug molecules in the fiber volume (D_f) and its diffusion transport in the interfibrillar space filled with a solvent (D_w). Consequently, the total transfer can be approximated by the model of a pseudo-two-layer medium. Following the models of Crank [76] and Mackey-Mears [77], the relationship between the effective diffusion coefficient (D_{eff}), individual diffusion constants (D_f, D_w) and the geometric characteristics of the system (L_M, R_f and L_w) has the form:

$$L_M/2D_{eff} = R_f/D_f + L_w/2D_w \quad (3)$$

here R_f and L_w are the average characteristic dimensions of the length of the diffusion path of the drug in the fiber and interfiber space, respectively, and L_M is the thickness of the fibrillar film.

Taking into account the symmetry of the film during double-sided desorption, its size and fiber diameter are divided by 2. For a cylindrical fiber, the length of the diffusion path is determined by its radius, while for L_w, we used a correction for the increase in the diffusion path due to the bending of the drug molecule around randomly arranged cylindrical fibers. The correction for impenetrable obstacles was first introduced in the classical work of Mackey and Mears [77] and was recently used to describe transport in a magnetic composite based on chitosan and PHB [78]:

$$L_w = [(1+\varphi_f)/(1-\varphi_f)]2L_M \quad (4)$$

where φ_f is the volume fraction of polymer fibers in the fibrillar film, the values of which, as well as the average radii, are given earlier in **Table 1**.

Taking into account the introduced correlation of the diffusion path length in the aqueous phase, and also taking into account the correction for the degree of

[Zn-TFP] %	τ 10^{10} s		τ 10^{10} s, annealed at140 ^{0}C		α_1/α_2,* %	ΔH_M, J/g	α_c, %
	Dense area	**Amorphous area**	**Dense area**	**Amorphous area**			
0	80.7	5.7	73	6	3.1	86	61
1	107	6	57	7.5	6.2	93	65
3	117	6.1	59	7.9	11	94	66
5	125	6.9	58	8.1	19	96	68

Subscripts 1 and 2 refer, respectively, to more dense (paracrystalline) and less dense (amorphous) regions of the intercrystalline space. ΔH_M—specific heat of fusion and α_c—degree of crystallinity of PHB.

Table 1.
Effective correlation time (τ) and volume ratio of paracrystalline (dense) ($\alpha 1$) and amorphous ($\alpha 2$) regions in an ultrathin PHB fiber containing Zn-TFP.

crystallinity of the polymer Fc = $(1-\alpha_c)$, proposed in monograph [79], Eq. (4) takes a more detailed form:

$$L_M/D_{eff} = R_f/([1-\alpha_c]D_f) + 2\,[(1+\varphi_f)/(1-\varphi_f)]\,L_M/D_w, \tag{5}$$

allowing you to go to the comparison of the contribution of two processes that determine the total transfer process, namely, drug diffusion in the fiber and diffusion in the aqueous interfibrillar space of the PHB film. A preliminary estimate of the values of the diffusion coefficients of the drug in PHB (D_f) showed that they are several orders of magnitude lower than the corresponding coefficient in the aqueous interfibrillar space, the last term on the right-hand side of Eq. (5) becomes much less than the first, which leads to a simplification of the sought expression for the diffusion coefficient of the drug in the hospital:

$$D_f = R_f D_{eff}/L_M Fc \tag{6}$$

where the geometric (φ_f, L_M) and structural (Fc = $1-\alpha_c$) characteristics were determined using the SEM and differential scanning calorimetry (DSC) data, respectively.

Simplification of Eq. (5) to expression (6) is possible only if the limiting stage of drug desorption from a fibrillar film of thickness LM is its diffusion in the polymer medium of the fiber. This is not universal, since with an increase in the volume of the interfibrillar space of the system, i.e. with a decrease in the proportion of fiber and its less dense packing, as well as with an increase in the concentration of hydrophilic groups in the polymer, i.e. with an increase in their affinity for a polar solvent (for example, for water) and a corresponding increase in Df, it is possible to take into account both terms, the values of which can be quite comparable.

The diffusion equation for an infinite cylinder which satisfactorily approximates the shape of the fiber and provided that the drug at the initial moment is uniformly distributed over the volume of the polymer was presented by Crank in his classic work [76]:

$$\partial G_d/\partial t = (1/r)D_f\,[\partial(r\partial G_d/\partial r)/\partial r]\;0 < r < R_f, \tag{7}$$

where r is the radial coordinate of diffusion, and the symbol G_d, as in Eq. (2), denotes the concentration of the mobile fraction of the drug in a cylindrical fiber with a corresponding constant diffusion coefficient D_f and R_f as before denotes the average radius of the fiber. The initial and boundary conditions corresponding to drug desorption from a cylindrical fiber are as follows: $G_d = G_{0d}$ at $t = 0$ (at the initial moment of time) and under the condition $0 < x < Rf$:

$G_d = 0$ at $r = R$ (at the fiber-solution interface) and under the condition $t > 0$.

The second boundary condition is written from considerations of symmetry and indicates the absence of flux at the center of a single fiber $\partial G_d/\partial r = 0$ at $r = 0$.

It was shown in [80] that the solution of Eq. (7) has the form of a power function and makes it possible to obtain the dependence of the cumulative amount of pharmaceuticals coming from the fiber into the environment by the diffusion mechanism $\Delta M_D(t)$ on its contact time with this medium t:

$$\Delta M_D/\Delta M_{D\infty} = \left[16D_f/\pi R_f^2\right]^{1/2} t^{1/2} - \left[D_f/R_f^2\right] t, \qquad (8)$$

where $\Delta M_{D\infty}$ is the limiting value of ΔM_D under the condition $t \to \infty$.

Note that Eqs. (7) and (8), describing the diffusion of drug through the side walls of an infinite cylinder, are valid provided that the length of the fiber exceeds its radius by at least five times [81], and this requirement is strictly fulfilled for fibers of practically infinite length, obtained by the technology of electrospinning. In addition, the last equation is valid provided that the inequality $\Delta M_D/\Delta M_{D\infty} < 0.4$. The combination of Eqs. (2) and (8) gives the final expression for the release of the drug from the cylindrical fibers, both taking into account the weight loss during hydrolysis and due to the diffusion of the drug:

$$\Delta M_d/\Delta M_{d\infty} = \left[16D_f/\pi R_f^2\right]^{1/2} t^{1/2} + k_c t, \qquad (9)$$

where $k_c = k_h - [D_f/R_f^2]$. The positive sign in the equation shows that under the given conditions for PHB, the inequality $k_h > [D_f/R_f^2]$ is fulfilled.

The experimental curves shown in **Figure 4** correspond to Eq. (9), which allows the calculation of the fiber diffusion coefficients D_f. For calculations, these drug release curves were presented in diffusion coordinates $\Delta M_D/\Delta M_{D\infty} \sim t^{1/2}$, as shown in **Figure 5**. Comparison of the experimental results presented in **Figure 3** with the corresponding symbols, and calculations according to Eqs. (7) and (9), shown by solid lines, indicates their good agreement and demonstrates the possibility of using this model to assess the diffusion characteristics of ultrathin cylindrical fibers. **Table 2** exhibits the concentration of mobile C^D_∞ and immobilized C^h_∞ drug fractions.

The effect of a sharp drug release at the initial portion of the kinetic curves is called the "burst effect" [82]. In our case, it is mainly associated with the residual drug concentration displaced from the fiber volume during the electrospinning process. The appearance of this fraction does not exceed 10% and is observed only for ultrafine fibers with a high drug loading (> 3%). When the sample is immersed in an aqueous medium, it is removed from the film due to rapid diffusion mobility in the hydrated interfibrillar space. It should also be noted that the concentration of mobile drug molecules decreases with an increase in crystallinity and therefore reaches a maximum value for fibers containing 1% DPD, having the lowest crystallinity of 38%. The last line of **Table 2** shows the constants of the drug yield due to the loss of fiber mass during hydrolysis (k_h). It can also be noted here that the rate of hydrolysis decreases with an increase in the crystallinity of the fiber and the lower, the higher the concentration of the drug in the fiber.

The totality of the results known in the literature and presented in this chapter allows us to draw two conclusions. First, the intercrystalline regions of ultrathin PHB

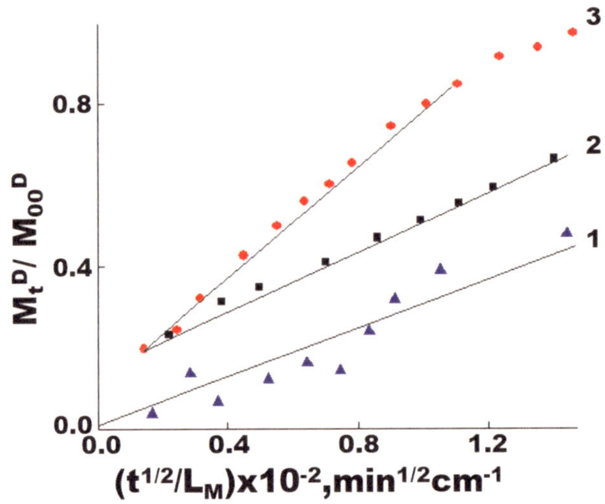

Figure 5.
Kinetic curves of DPD release from PHB fibers, presented in diffusion coordinates. The concentrations of DPD are 1 (1), 3 (2), and 5 wt.% (3); L_M is the total thickness of the fibrillar film.

DPD, wt %	$R_f \times 10^4$, cm^2	$D_f \times 10^{12}$, cm^2/s	$C^D_\infty \times 10^2$, g/g	$C^h_\infty \times 10^2$, g/g	$k_c \times 10^5$, s^{-1}	$k_h \times 10^5$, s^{-1}
5	1.6	1.5	3.35 (53%)	1.65 (34%)	4.4	1.4
3	2.5	3.2	1.8 (56%)	0.89 (32%)	6.1	1.1
1	9.8*	6.9*	0.78 (78%)	0.22 (22%)	9.4	8.5*

The conditional calculations of the corresponding characteristics, calculated approximating the ellipsoids of revolution with cylinders of larger diameter.

Table 2.
Sorption, kinetic, and diffusion characteristics of the PHB-DPD fiber system. Concentration of mobile C^D_∞ and fixed C^h_∞ fractions of DPD in fiber; C^{Ext}—concentration of DPD in the interfiber space. The percentages of each fraction are indicated in parentheses.

fibers have a close packing of chains, which is significantly higher than the density in the amorphous region of the film, which is confirmed by EPR measuring. Second, under the same conditions, the values of drug diffusion coefficients in ultrathin and highly crystalline PHB fibers, as well as in its spherical microparticles containing DPD [83], are several orders of magnitude lower than similar characteristics measured for PHB films.

6. Conclusion

Over the past 15 years, there has been not only significant progress at the level of theory and modeling of the processes of directed transport of biologically active compounds, but at the same time a large number of polymer and hybrid materials were created for modern therapy, providing an innovative component for drug delivery. Today, the main trend is the accumulation of experimental data showing how nanotechnological processes, including electrospinning, affect the characteristics of the developed means of targeted drug delivery.

Author details

Anatoly A. Olkhov[1,2], Svetlana G. Karpova[3], Anna V. Bychkova[3], Alexandre A. Vetcher[4,5]* and Alexey L. Iordanskii[1]

1 N.N. Semenov Federal Research Center for Chemical Physics, Russian Academy of Sciences, Moscow, Russia

2 Plekhanov Russian University of Economics, Moscow, Russia

3 Emanuel Institute of Biochemical Physics of Russian Academy of Sciences, Moscow, Russia

4 Peoples' Friendship University of Russia (RUDN), Moscow, Russia

5 Complementary and Integrative Health Clinic of Dr. Shishonin, Moscow, Russia

*Address all correspondence to: avetcher@gmail.com

References

[1] Ariga K, Vinu A, Miyahara M. Recent progresses in bio-inorganic nanohybrids. Current Nanoscience. 2006;**2**(3):197

[2] Colilla M, Manzano M, Vallet-Ragí M. Recent advances in ceramic implants as drug delivery systems for biomedical applications. International Journal of Nanomedicine. 2013;**3**(4):403

[3] Dubey KA, Chaudhari CV, Bhardwaj YK, Varshney L. Polymers, blends and nanocomposites for implants, scaffolds and controlled drug release applications. Advanced Structured Materials. 2017;**66**:1

[4] Hernández-Gómez LH, Beltrán-Fernández JA, Ramírez-Jarquín M, Pava-Chipol N, Pérez-Montejo S. Characterization of scaffold structures for the development of prostheses and biocompatible materials. Advanced Structured Materials. 2019; **92**:471

[5] Mann JL, Yu AC, Agmon G, Appel EA. Supramolecular polymeric biomaterials. Biomaterials Science. 2018; **6**(1):10

[6] LaVan DA, McGuire T, Langer R. Small-scale systems for in vivo drug delivery. Nature Biotechnology. 2003; **21**(10):1184

[7] Ishihara T, Mizushima T. Techniques for efficient entrapment of pharmaceuticals in biodegradable solid micro/nanoparticles. Expert Opinion on Drug Delivery. 2010;7(5):565

[8] Haidar MK, Eroğlu H. Nanofibers: New insights for drug delivery and tissue Engineering. Current Topics in Medicinal Chemistry. 2017;**17**(13): 1564

[9] Bhardwaj N, Kundu SC. Electrospinning: A fascinating fiber fabrication technique. Biotechnology Advances. 2010;**28**(3):325

[10] Zhang Y, Chen Q, Ge J, Liu Z. Controlled display of enzyme activity with a stretchable hydrogel. Chemical Communications. 2013;**49**:84

[11] Park S, Lee Y, Kim DN, Park S, Jang E, Koh WG. Entrapment of enzyme-linked magnetic nanoparticles within poly(ethylene glycol) hydrogel microparticles prepared by photopatterning. Reactive and Functional Polymers. 2009;**69**(5):293. DOI: 10.1016/j.reactfunctpolym.2009. 02.001

[12] Thornton PD, Mart RJ, Ulijn RV. Enzyme-responsive polymer hydrogen hydrogel particles for controlled release. Advanced Materials. 2007;**19**(9):1252

[13] Gupta B, Revagade N, Hilborn J. Poly (lactic acid) fiber: An overview. Progress in Polymer Science. 2007;**32**(4):455

[14] Zhang B, Yan X, He H-W, Ning X, Long Y-Z. Solvent-free electro-spinning: Opportunities and challenges. Polymer Chemistry. 2017;**8**(2):334

[15] Streicher RM, Schmidt M, Fiorito S. Nanosurfaces and nanostructures for artificial orthopedic implants. Nanomedicine. 2007;**2**(6):861

[16] Li D, Frey MW, Baeumner AJ. Electrospun polylactic acid nanofiber membranes as substrates for biosensor assemblies. Journal of Membrane Science. 2006;**279**(1-2):354

[17] Mikheev AY, Shlyapnikov YM, Kanev IL, Avseenko AV, Morozov VN.

Filtering and optical properties of free standing electrospun nanomats from nylon-4,6. European Polymer Journal. 2016;**75**:317

[18] Miguel SP, Figueira DR, Simões D, Ferreira P, Correia IJ. Electrospun polymeric nanofibres as wound dressings: A review. Colloids and Surfaces B: Biointerfaces. 2018;**169**:60

[19] Zhou B, Li Y, Deng H, Hu Y, Li B. Antibacterial multilayer films fabricated by layer-by-layer immobilizing lysozyme and gold nanoparticles on nanofibers. Colloids and Surfaces B: Biointerfaces. 2014;**116**:432

[20] Cheng H, Yang X, Che X, Yang M, Zhai G. Biomedical application and controlled drug release of electrospun fibrous materials. Materials Science and Engineering. 2018;**90**:750

[21] Rebia RA, Rozet S, Tamada Y, Tanaka T. Biodegradable PHBH/PVA blend nanofibers: Fabrication, characterization, in vitro degradation, and in vitro biocompatibility. Polymer Degradation and Stability. 2018;**154**:124

[22] Malafeev KV, Moskalyuk OA, Yudin VE, Popova EN, Ivan'kova EM. Synthesis and properties of fibers prepared from lactic acid–glycolic acid copolymer. Polymer Science, Series A. 2017;**59**(1):53

[23] Cao K, Liu Y, Olkhov AA, Siracusan V, Iordanskii AL. PLLA-PHB fiber membranes obtained by solvent-free electrospinning for short-time drug delivery. Drug Delivery and Translational Research. 2018;**8**:291

[24] Bose S, Tarafder S. Calcium phosphate ceramic systems in growth factor and drug delivery for bone tissue engineering: A review. Acta Biomaterialia. 2012;**8**(4):1401

[25] Sokolova VV, Radtke I, Heumann R, Epple M. Effective transfection of cells with multi-shell calcium phosphate-DNA nanoparticles. Biomaterials. 2006; **27**(16):3147

[26] Dorati R, DeTrizio A, Modena T, et al. Biodegradable scaffolds for bone regeneration combined with drug-delivery systems in osteomyelitis therapy. Pharmaceuticals. 2017;**10**(4): 96. DOI: 10.3390/ph10040096

[27] Nandi SK, Bandyopadhyay S, Das P, Samanta I, Mukherjee P, Roy S, et al. Understanding osteomyelitis and its treatment through local drug delivery system. Biotechnology Advances. 2016; **34**(8):1305

[28] Das S, Baker AB. Biomaterials and nanotherapeutics for enhancing skin wound healing. Frontiers in Bioengineering and Biotechnology. 2016; **4**:82

[29] Tran CD, Duri S, Harkins AL. Recyclable synthesis, characterization and antimicrobial activity of chitozan-based polysaccharide composite materials. Journal of Biomedical Materials Research Part A. 2013;**0**(8): 2248

[30] Tan H, Chu CR, Payne K, Marra KG. Injectable in situ forming biodegradable chitosan-hyaluronic acid based hydrogels for cartilage tissue engineering. Biomaterials. 2009;**30**(13): 2499. DOI: 10.1016/j.biomaterials.2008.12.080

[31] Song Y, Nagai N, Saijo S, Kaji H, Nishizawa M, Abe T. In situ formation of injectable chitosan-gelatin hydrogels through double crosslinking for sustained intraocular drug delivery. Materials Science & Engineering. C, Materials for Biological Applications. 2018;**1**(88):1

[32] Vikhoreva G, Bannikova G, Stolbushkina P, Panov A, Drozd N, Makarov V, et al. Preparation and anticoagulant activity of a low-molecular-weight sulfated chitosan. Carbohydrate Polymers. 2005;**62**:327

[33] Xue J, Zhao W, Nie S, Sun S, Zhao C. Blood compatibility of polyethersulfone membrane by blending a sulfated derivative of chitosan. Carbohydrate Polymers. 2013;**95**:64. DOI: 10.1016/j.carbpol.2013.02.033

[34] Huang X, Wang R, Lu T, Zhou D, Zhao W, Sun S, et al. Heparin-like chitosan hydrogels with tunable swelling behavior, prolonged clotting times, and prevented contact activation and complement activation. Biomacromolecules. 2016;**17**:4011. DOI: 10.1021/acs.biomac.6b01386

[35] Zhang L, Ma Y, Pan X, Chen S, Zhuang H, Wang S. A composite hydrogel of chitosan/heparin/poly (γ-glutamic acid) loaded with superoxide dismutase for wound healing. Carbohydrate Polymers. 2018;**180**:168

[36] Qiao X, Peng X, Qiao J, Jiang Z, Han B, Yang C, et al. Evaluation of a photocrosslinkable hydroxyethyl chitosan hydrogel as a potential drug release system for glaucoma surgery. Journal of Materials Science. Materials in Medicine. 2017;**28**:149

[37] Nguyen HX, O'Leary EA. An in vitro thrombolysis study using a mixture of fast-acting and slower release microspheres. Pharmaceutical Research. 2016;**33**:1552. DOI: 10.1007/s11095-016-1897-1

[38] Dobiasch S, Szanyi S, Kjaev A, Werner J, Strauss A, Weis C, et al. Synthesis and functionalization of protease-activated nanoparticles with tissue plasminogen activator peptides as targeting moiety and diagnostic tool for pancreatic cancer. Journal of Nanobiotechnology. 2016;**14**:81. DOI: 10.1186/s12951-016-0236-3

[39] Zhang LL, Li P, Li YM, Wang AQ. Preparation and characterization of magnetic alginate-chitosan hydrogel beads loaded matrine. Drug Development and Industrial Pharmacy. 2012;**38**(7):872. DOI: 10.3109/03639045.2011.630397

[40] Zhang Y, Sun Y, Yang X, Hilborn J, Heerschap A, Ossipov DA. Injectable in situ forming hybrid iron oxide-hyaluronic acid hydrogel for magnetic resonance imaging and drug delivery. Macromolecular Bioscience. 2014;**14**(9):1249

[41] Iordanskii AL, Rogovina SZ, Berlin AA. Current state and developmental prospects for nanopatterned implants containing drugs. Review Journal of Chemistry. 2013;**3**(2):117-132

[42] Leoni L, Desai TA. Nanoporous biocapsules for the encapsulation of insulinoma cells: Biotransport and biocompatibility considerations. IEEE Transactions on Biomedical Engineering. 2001;**48**(11):1335

[43] Mandracchia D, Tripodo G. Chapter 1: Micro and nano-drug delivery systems. In: Silk-based Drug Delivery Systems. 2020. pp. 1-24. DOI: 10.1039/9781839162664-00001. eISBN: 978-1-83916-266-4

[44] Rajangam T, An SS. Fibrinogen and fibrin based micro and nano scaffolds incorporated with drugs, proteins, cells and genes for therapeutic biomedical applications. International Journal of Nanomedicine. 2013;**8**:3641

[45] Ziv-Polat O, Skaat H, Shahar A, Margel S. Novel magnetic fibrin hydrogel scaffolds containing thrombin and growth factors conjugated iron oxide nanoparticles for tissue engineering. International Journal of Nanomedicine. 2012;7:1259

[46] Agapova OI. Silk fibroin and spidroin bioengineering constructions for regenerative medicine and tissue engineering (review). Sovremennye Tehnologii v Medicine (Modern Technologies in Medicine). 2017;**9**(2): 190-206

[47] Gomzyak VI, Demina VA, Razuvaeva EV, Sedush NG, Chvalun SN. Biodegradable polymer materials for medical applications: From implants to organs. Fine Chemical Technologies. 2017;**12**(5):5-20. (In Russ.)

[48] Kinard LA, Kasper FK, Mikos AG. Synthesis of oligo(poly(ethylene glycol) fumarate). Nature Protocols. 2012;7(6): 1219

[49] Santoro M, Tatara AM, Mikos AG. Gelatin carriers for drug and cell delivery in tissue engineering. Journal of Controlled Release. 2014;**0**:210. DOI: 10.1016/j.jconrel.2014.04.014

[50] Modaresi SM, Mehr SE, Faramarzi MA, Gharehdaghi EE, Azarnia M, Modarressi MH, et al. Preparation and characterization of self-assembled chitosan nanoparticles for the sustained delivery of streptokinase: An in vivo study. Pharmaceutical Development and Technology. 2014;**19**: 593. DOI: 10.3109/10837450.2013.813542

[51] Jin HJ, Zhang H, Sun ML, Zhang BG, Zhang JW. Urokinase-coated chitosan nanoparticles for thrombolytic therapy: Preparation and pharmacodynamics in vivo. Journal of Thrombosis and Thrombolysis. 2013;**36**:458

[52] Mourino V, Boccaccini AR. Bone tissue engineering therapeutics: Controlled drug delivery in three-dimensional scaffolds. Journal of the Royal Society Interface. 2010;7(43):209. DOI: 10.1098/rsif.2009.0379

[53] Vatchha SP, Kohli A, Tripathi SK, Nanda SN, Pradhan P, Shiraz SM. Biodegradable implants in orthopaedics (review). Annals of International Medical and Dental Research. 2015;**1**(1): 3-8

[54] Zhao F, Yao D, Guo R, Deng L, Dong A, Zhang J. Composites of polymer hydrogels and nanoparticulate systems for biomedical and pharmaceutical applications. Nanomaterials. 2015;**5**(4): 2054. DOI: 10.3390/nano5042054

[55] Bukreeva TV, Sulyanov SN, Korotkov NY, Rumyantseva SS, Marchenko IV, Funtov KO, et al. In situ synthesis and characterization of magnetic nanoparticles in shells of biodegradable polyelectrolyte microcapsules. Materials Science and Engineering. 2014;**45**:225

[56] Albuquerque PBS, Malafaia CB. Perspectives on the production, structural characteristics and potential applications of bioplastics derived from polyhydroxyalkanoates. International Journal of Biological Macromolecules. 2018;**107**:615

[57] Sevastianov VI, Perova NV, Shishatskaya EI, Kalacheva GS, Volova TG. Production of purified polyhydroxyalkanoates (PHAs) for applications in contact with blood. Journal of Biomaterials Science, Polymer Edition. 2003;**14**:1029

[58] Artsis MI, Bonartsev AP, Iordanskii AL, Bonartseva GA, Zaikov GE. Biodegradation and medical application of microbial

poly(3-hydroxybutyrate). Molecular Crystals and Liquid Crystals. 2010;**523**: 593. DOI: 10.1080/15421401003726519

[59] Ivantsova EL, Kosenko RY, Iordanskii AL, Rogovina SZ, Prut EV, Filatova AG, et al. Structure and prolonged transport in a biodegradable poly(r-3-hydroxybutyrate)-drug system. Polymer Science, Series A. 2012;**54**(2): 87-93

[60] Karpova SG, Popov AA, Lomakin SM, Shilkina NG, Iordanskii AL, Klenina NS, et al. Changes in the structural parameters and molecular dynamics of polyhydroxybutyrate-chitosan mixed compositions under external influences. Russian Journal of Physical Chemistry B. 2013;**7**(3):225-231

[61] Olkhov AA, Karpova SG, Staroverova OV, Kucherenko EL, Ishchenko AA, Iordanskii AL. Effect of external factors on the structure of ultrathin fibers of poly(3-Hydroxybutyrate) and dipyridamole. Fibre Chemistry. 2017;**48**(4):1

[62] Karpova SG, Olkhov AA, Shilkina NG, Iordanskii AL. Investigation of biodegradable composites of poly(3-hydroxybutyrate) ultrathin fibers modified by a complex of iron(III) with tetraphenylporphyrin. Polymer Science, Series A. 2017;**59**(3): 343. DOI: 10.1134/S0965545X17030075

[63] Cho D, Zhmayev E, Joo YL. Structural studies of electrospun nylon 6 fibers from solution and melt. Polymer. 2011;**52**:4600

[64] Vasserman AM, Buchachenko AL, Kovarskii AL, Neiman MB. Study of molecular motion in polymers by the paramagnetic probe method. Polymer Science U.S.S.R. 1968;**10**(8): 2238. DOI: 10.1016/0032-3950(68)90317-1

[65] Di Lorenzo ML, Gazzano M, Righetti MC. The role of the rigid amorphous fraction on cold crystallization of poly(3-hydroxybutyrate). Macromolecules. 2012;**45**(14):5684. DOI: 10.1021/ma3010907

[66] Kamaev PP, Aliev II, Iordanskii AL, Wasserman AM. Molecular dynamics of the spin probes in dry and wet poly(3-hydroxybutyrate) films with different morphology. Polymer. 2001;**42**(2):515. DOI: 10.1016/S0032-3861(00)00339-6

[67] Ozerin AN, Zurov YA, Selikhova VI, Chvalun SN, Bakeyev NF. Methods of investigation use of the method of measuring the absolute intensity of small angle x-ray scattering for the study of the structure of amorphous regions in oriented polyethylene films. Polymer Science U.S.S.R. 1976;**18**(9):2434-2442

[68] Huynh CT, Kang SW, Li Y, Kim BS, Lee DS. Controlled release of human growth hormone from a biodegradable pH/temperature-sensitive hydrogel system. Soft Matter. 2011;**7**(19):8984

[69] Hu C, Liu S, Zhang Y, Li B, Yang H, Fan C, et al. Long-term drug release from electrospun fibers for in vivo inflammation prevention in the prevention of peritendinous adhesions. Acta Biomaterialia. 2013;**9**:7381

[70] Padmanabhan J, Kyriakides TR. Nanomaterials, inflammation, and tissue engineering. Wiley Interdisciplinary Reviews: Nanomedicine and Nanobiotechnology;**7**(3):355

[71] Keshavarz P, Ayatollahi S, Fathikalajahi J. Mathematical modeling of gas–liquid membrane contactors using random distribution of fibers. Journal of Membrane Science. 2008; **325**:98

[72] Petlin DG, Amarah AA, Tverdokhlebov SI, Anissimov YG. A fiber distribution model for predicting drug release rates. Journal of Controlled Release. 2017;**258**:218

[73] Nakielski P, Kowalczyk T, Zembrzycki K, Kowalewski TA. Experimental and numerical evaluation of drug release from nanofiber mats to brain tissue. Journal of Biomedical Materials Research—Part B: Applied Biomaterials. 2015;**103**(2):282

[74] Iordanskii AL, Rogovina SZ, Kosenko RY, Ivantsova EL, Prut EV. Development of a biodegradable polyhydroxybutyrate-chitosan-rifampicin composition for controlled transport of biologically active compounds. Doklady Physical Chemistry. 2010;**431**(2):60-62. DOI: 10.1134/S0012501610040020

[75] Sevim K, Pan J. A model for hydrolytic degradation and erosion of biodegradable polymers. Acta Biomaterialia. 2018;**66**:192-199. DOI: 10.1016/j.actbio.2017.11.023

[76] Crank J. The Mathematics of Diffusion. Oxford: Clarendon Press, 1992.

[77] Mackie JS, Meares P. The diffusion of electrolytes in a cation-exchange resin membrane I. Theoretical. Proceedings of the Royal Society of London. Series A. 1995;**232**(1191):498. DOI: 10.1098/rspa.1955.0234

[78] Bychkova AV, Iordanskii AL, Kovarski AL, Sorokina ON, Kosenko RY, Markin VS, et al. Nanotechnologies in Russia. 2015;**10**(3–4):325

[79] Mikhailov YM, Ganina LV, Shapaeva NV, Chalykh AE. Diffusion of Low-molecular mass substances in the region of polymer-glass transition. Polymer Science, Series B. 1999;**41**(5-6): 120-123

[80] Siepmann J, Siepmann F. Mathematical modeling of drug delivery. International Journal of Pharmaceutics. 2008;**364**:328

[81] Jamstorr BE. Acta Universitatis Upsaliensis. Vol. 884. Uppsala: Digital Comprehensive Summaries of Uppsala Dissertations from the Faculty of Science and Technology. p. 79

[82] Huang X, Brazel CS. On the importance and mechanisms of burst release in matrix-controlled drug delivery systems. Journal of Controlled Release. 2001;**73**(2–3):121. DOI: 10.1016/S0168-3659(01)00248-6

[83] Bonartsev AP, Livshits VA, Makhina TA, Myshkina VL, Bonartseva GA, Iordanskii AL. Controlled release profiles of dipyridamole from biodegradable microspheres on the base of poly(3-hydroxybutyrate). Express Polymer Letters. 2007;**1**(12):797

Chapter 6

Functional Nanofibers for Sensors

Stanislav Petrík and Mayza Ibrahim

Abstract

Electrospun nanomaterials and their applications have increasingly gained interest over the last decade. Nanofibers are known for their exceptional surface area and wide opportunities for their functionalization. These properties have been attractive for various sensing applications; however, mostly electric sensing principles have been reported. An overview of most frequently studied concepts will be presented. A novel approach based on optical detection will be described. Various functionalized nanofiber materials have been used to demonstrate feasibility of realization of miniature sensors of biomedical and chemical values (enzymes reactions, metal ions content, concentration, etc.). Compactness and sensitivity of the sensors are significantly enhanced through original hybrid fiber-optic/nanofiber design. The potential of the new detection principle for various applications (bio-medical, chemical, forensic, automotive, etc.) will be discussed.

Keywords: electrospinning, functional nanofibers, conductive composite nanofibers, inorganic semi-conductive nanofibers, sensors, optical fiber

1. Introduction

Electrospinning is a highly versatile technique to produce continuous fibers with diameters ranging from several micrometers down to few nanometers by applying a high voltage on a solution or melts, mainly from polymers. At nanoscale, several superior characteristics occur such as large surface to volume ratio that can reach values as large as 10^3 times of that of micrometer, easy adaptability to surface functionalization, and extraordinary supreme mechanical properties such as stiffness and tensile strength. These outstanding characteristics make electrospun nanofibers an optimal candidate for many important applications [1].

Beside electrospinning, a number of processing methods have been used in recent years to produce polymer nanofibers, such as drawing, self-assembly, template synthesis, and phase separation [2]. Each of these techniques has its limitation, whereas drawing is only limited to viscoelastic materials that can handle the stresses developed during pulling to produce nanofibers, while self-assembly is time consuming in producing continuous polymer nanofibers. Template synthesis uses nanoporous membrane as a template to produce nanofibers of solid (a fibril) or hollow (a tubule) shape. Phase separation takes relatively long time to transfer solid polymer into nanoporous foam. Electrospinning process due to its ease of fabrication appears to be the only technique, which could be further developed for mass production.

The term electrospinning has been used relatively recently; however, its fundamentals dated back more than 60 years earlier. Formhals published a succession of patents [1, 3–6] from 1934 to 1944. Through this series, he specified the experimental setup for producing polymer filaments using electrostatic force, whereas the polymer solution was exposed to electric field through two electrodes with different polarity. One is placed into the solution, and the other onto the collector. Once the jet solution ejected out from a metal spinneret, it evaporated to become fibers and these fibers were collected on the collector. The potential difference depended on the properties of the solution such as polymer molecular weight and viscosity. The problem occurred that was the fibers favored to stick to each other as well as to the collectors. This problem was due to the insufficient distance between the spinneret aperture and the collectors, which led to inadequate time for jet solution to evaporate. In 1936, C.L. Norton approach was patented due to his contribution to electrospinning from a melt rather than solution using air blast to boost fiber formation [7], Rozenblum and Petryanov-Sokolov [8] in 1938 produced electrospun fiber that was developed into filter materials. These filter materials were then mass manufactured for gas masks. Sir Geoffrey Ingram Taylor stablished the underpinning of a theory for electrospinning between 1964 and 1969. He explained the mathematical model of the cone shape of the fluid droplet under the electric field [9–11]. In the report of the National Institutes of Health (NIH), The Small Business Innovation Research 1988, Simon produces a submicron- and nanoscale fibrous mats from electrospinning. These mats were especially created for use as substrates *in vitro* cell [12]. In the beginning of 1990, many organic polymers have been successively elctrospun into nanofibers. Credit for that goes to many research groups, remarkably Reneker and Rutledge, who familiarize the name electrospinning for the process. This process can be simply explained when a sufficiently high voltage is applied to a liquid droplet, and it will charge the body of

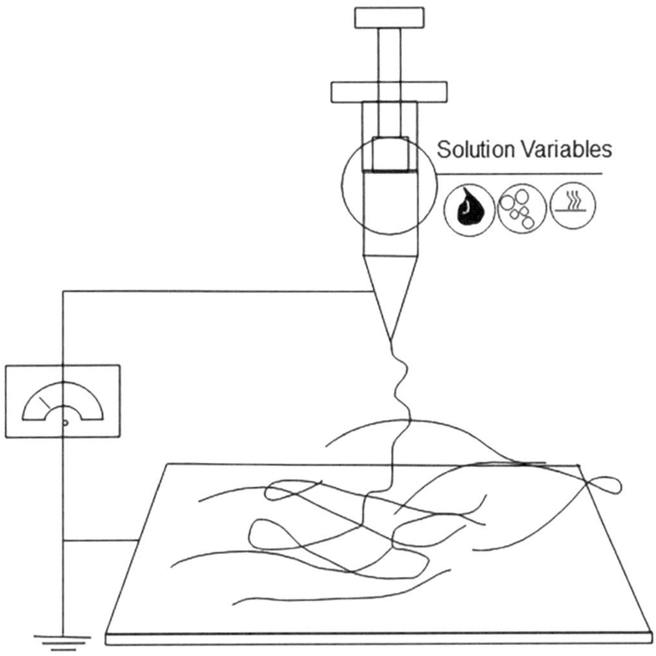

Figure 1.
Electrospinning apparatus schematic.

the droplet. The electrostatic repulsion will generate to counteract the surface tension. Hence, the droplet will stretch. At critical point named Taylor zone, a stream of liquid will be erupted from the surface. If the molecular cohesion of the liquid is sufficiently high, a charged jet will be formed; otherwise, droplets are electrosprays. As the jet flies in air, it will eventually dry and will deposit on the grounded collector.

The standard laboratory setup for electrospinning apparatus consists of spinneret connected to high-voltage (5–50 kV) direct current power supply as illustrated in **Figure 1**.

There is a variety of solutions that can be loaded into the syringe, for example, polymer solution, sol–gel, particulate suspension, or melt [13].

By controlling the processing parameters, different nanofiber morphologies can be obtained (**Figure 2a-m** [14]), beaded, smooth [15], helical [16], ribbon [17], necklace-like [18], porous [18], core-shell [19], hollow [20], multichannel-tubular [21], nanowire-in microtube [22], muli-core cable-like [23], tube-in-tube structured nanofibers [24].

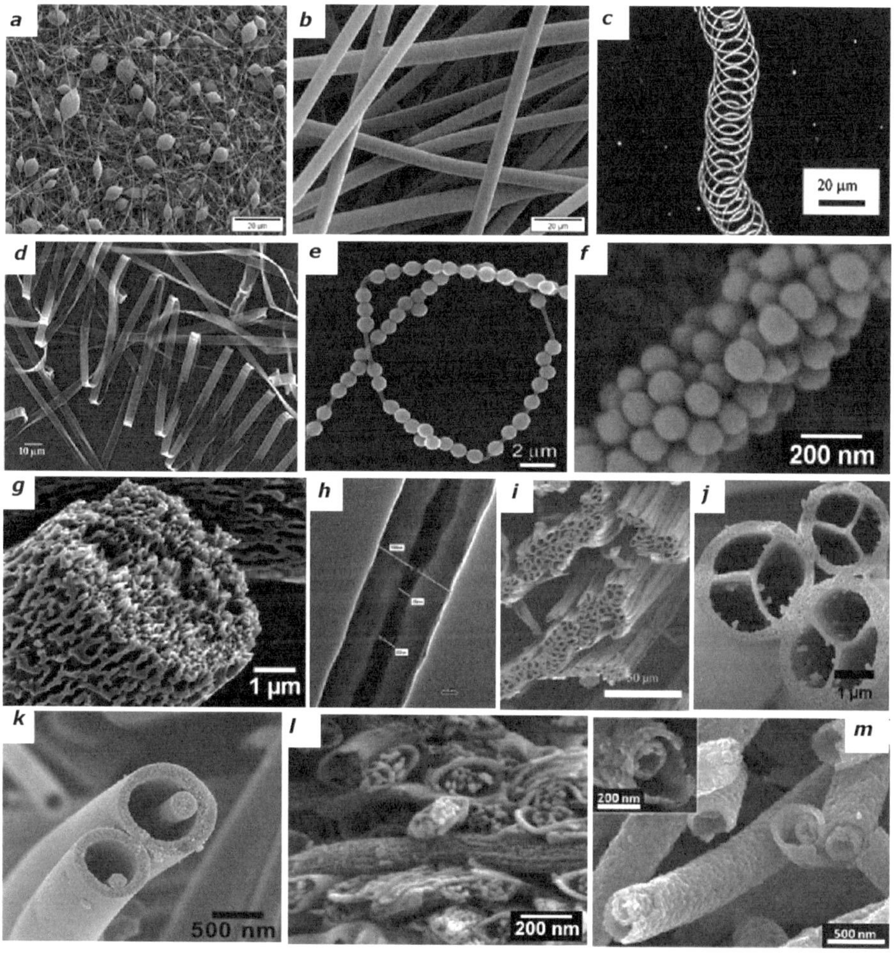

Figure 2.
Different nanofiber morphologies: (a) beaded (b) smooth, (c) helical, (d) ribbon, (e) necklace-like, (f,g) porous, (h) core-shell, (i) hollow, (j) multichannel-tubular, (k) nanowire-in microtube, (l) muli-core cable-like (m) tube-in-tube structured nanofibers.

2. Electrospinning and sensors

The main properties that should be provided in any material to be used as a sensor are: firstly, to be responsive to the external stimuli; secondly, this response can be accessible to electronic interface; thirdly, this has a large specific area since sensing preferentially occurs at interface. The material that possesses the first property is called smart material. The stimuli can be pressure, temperature, PH, moisture, chemical substances, electric, magnetic field, or light. In order for these smart materials to a sensor, it should act as a transducer. That means to response to the external stimuli in a way which can be measurable. In other words, it converts the external stimuli into a quantity that can be measurable.

Electrospinning manifests the capabilities of smart materials at the nanoscale dimension, especially as sensing materials. At nanoscale dimensions, many features are accessible, for example, excellent mechanical properties, especially flexibility, high porosity, large surface area, ability to surface functionality, and the ability to produce not only one-dimensional (1D), but also three-dimensional (3D) materials. Due to dramatic decrease in the diameter of the fibers, this has a great impact on the surface area, which is significantly increased, consequently the number of sites increase to interact with the external environment more effectively. High porosity provides utmost channels for transporting among nanofibers in electrospun mats, hence speeding up the transportation mechanism and increasing sensitivity. Another aspect of electrospinning is in its ability to form continuous nanofibers. This feature is so important in sensors, because sensors are usually assembled into a certain measuring instrument or analog-to-digital conversion circuit; therefore, it should provide a stable continuous circuit to supply a path for the current. Hence, electrospinning is irreplaceable to provide a stable circuit. Additionally, electrospinning makes use of various materials from inorganic to organic matters.

Smart materials, also named stimuli-responsive materials, are capable of undergoing reversible physical/chemical change upon exposure to external stimulus, such as temperature, PH, electrical, magnetic, light, chemicals, ions. Integration of these stimuli responsive materials with nanotechnology, such as electrospinning, has enormously accelerated the development of sensors.

3. Transduction sensing mechanisms

3.1 Capacitive sensors based on electrospun nanofibers

Capacitive sensors mainly depend on changing the relative permittivity of the dielectric material between two conducting electrodes. Capacitive devices are often used as displacement and pressure sensors.

Yang et al. [25] developed a flexible capacitive pressure sensor based on electrospun polyvinylidene fluoride (PVDF) nanofiber membrane with carbon nanotubes (CNTs). The fabrication process and schematic diagram of CNT-PVDF composite nanofiber are shown in **Figure 3a,b**. Two pieces of indium tin oxide polyethylene terephthalate films connected with cooper wires were fixed on the top and bottom surface of the composite nanofiber membrane as electrodes to record the capacitance variation under external pressure. The schematic diagram and the actual diagram of the single sensor are shown in **Figure 3c**. The SEM images of the composite nanofibers are shown in **Figure 3d,g** with different CNT weight ratios of 0.03, 0.05,

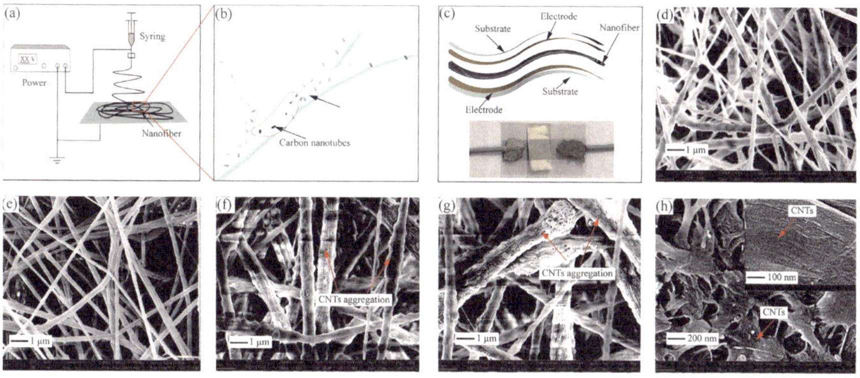

Figure 3.
Flexible capacitive pressure sensor: (a) schematic of the fabrication of the composite nanofiber. (b) schematic diagram of the composite nanofiber. (c) Diagram of the sensor. (d)–(h) SEM images of the composite nanofiber membrane with 0.03, 0.05, 0.1, 0.2 wt % carbon nanotube additions, respectively [25].

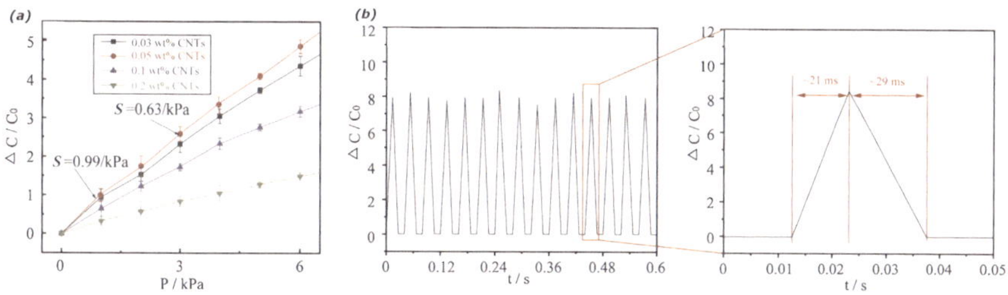

Figure 4.
Characterization of the pressure sensing performance of the flexible capacitive sensor. (a) The relative change in capacitance of the sensor with different weight ratio CNTs addition under low pressure applied. (b) Experimental systems for dynamic pressure applying and measuring capacitance, and enlarged view of a portion of the figure.

0.1, and 0.2 wt % of PVDF. At the beginning, the increase of CNT led to decrease the diameter of nanofibers, and then, the diameter increases again. This is because the when CNT increases the conductivity, the electrostatic force between the collector and the syringe spinneret will increase. Therefore, the nanofibers will be pulled thinner. Any further increase of CNT will lead to increase of nanofibers diameter as the CNTs tend to agglomerate due to strong Van der Waals force.

By increasing the permittivity and decreasing the young's modulus of the CNT-PVDF dielectric layer, the capacitive sensor exhibited high sensitivity (~0.99/kPa) with a composition of 0.05 wt% CNTs (**Figure 4a**), fast response (~29 ms), and excellent cyclic loading/unloading stability (>1000 cycles) (**Figure 4b**).

3.2 Resistive sensors based on electrospun nanofibers

Resistive sensors rely on the measuring the change of electrical resistivity as a variable of the amount of the reactive analytical samples through surface reaction. Up to now, numerous attempts have been carried on to develop ultrasensitive sensors to detect NH_3, CO_2, CO, O_2, H_2S, moisture, volatile organic compounds (VOCs) [26–30]. Resistive sensors based on electrospun nanofibers provide high and quick gas response *via* rapid and effective diffusion of analytic gas on the sensing surface. The porosity of the sensitive layer supply high-diffusivity paths for target molecules, promoting gas diffusion and mass transportation across sensing surface. This leads to faster response/recovery times. Moreover, porous surface boosts the formation of more chemisorbed oxygen species [31, 32].

3.2.1 Inorganic nanofibers

Resistive sensors based on metal oxide semiconductors (MOS) are the most simple and versatile gas sensors [33].

The resistance of the metal oxide is changing in accordance with the adsorption of the gases. One-dimensional (1D) nanostructured MOS have attracted much attention as chemical sensor materials as a result of their large surface-to-volume ratio, high porosity, excellent surface activities, and high surface charge modulation depth. ZnO, SnO_2, TiO_2, NiO, and $LaFeO_3$ are the most 1D MOS, which have been widely developed in the creation of highly sensitive gas sensors [34, 35].

The sensing mechanism of MOS gas sensors can be illustrated as follows:

In pure air, donor electrons in metal oxide attract to the oxygen, which is adsorbed into the surface of sensing material, preventing current flow (**Figure 5a**), while in the presence of the target gas (**Figure 5b**), oxygen reacts with the reducing gases. Hence, surface density of adsorbed oxygen decreases, and those electrons are then released into MOS, allowing current to flow freely through the sensor.

Abundance of n-type semiconductors such as ZnO, SnO_2, TiO_2, In_2O_3, WO_3, and ZnO/SnO_2 have turned out to be excellent gas materials for detecting both reducing and oxidizing gases, including H_2, NH_3, ethanol, acetone, and toluene. TiO_2 is the most well-known MOS used in ultrasensitive resistive sensors. Kim et al. have indicated the use of TiO_2 nanofibers as a detector for NO_2 [30]. Wang et al. reported that ZnO nanofibers with an average diameter of 150 nm display excellent sensing properties against ethanol at an operating temperature of 300°C, with a rapid response of about 3 s, including short recovery time of about 8 s and high sensitivity [36]. Lately, SnO_2 has attracted much attention because of its high transparency, semiconductivity, wide-band gap, and huge magneto-optic and chemical sensing effects [37, 38]. A highly porous SnO_2 nanofibers were prepared by combining electrospinning with oxygen plasma etching. They displayed fast response (7 s), wide linear response range, and low detection limit (< 1 ppb) [39].

In addition, doping is an efficient method to improve the sensing properties of the sensors. Li et al. have demonstrated that LiCl-doped TiO_2 nanofibers have an enhanced sensitivity toward humidity better than pure TiO_2 nanofibers [40].

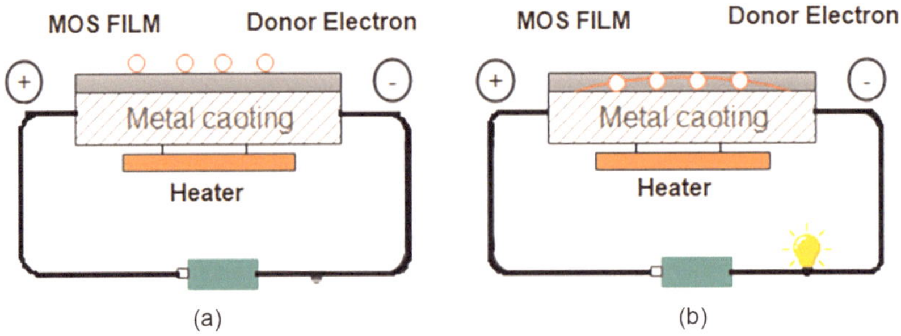

Figure 5.
Sensing mechanism of MOS gas sensors: (a) in clean air and (b) in the presence of the target gas.

Moreover, the composite nanofibers have ultra-fast response and recovery time. Zhang et al. developed double-layer ZnO/In_2O_3 composite nanfibers for sensing ethanol. $ZnO/In_2O_3/ZnO$ displayed improved and excellent sensing properties compared with ZnO nanofibers (detection limit of 1 ppm, shorter response, and recovery time of 2 and 1 s, respectively).

In addition to n-type semiconductors, p-type semiconductors have also been used to prepare vapor sensors, including NiO, Cr_2O_3, $LaFeO_3$, CuO, LaOCl/NiO, etc. Fan et al. [41] produced $LaFeO_3$ nanofibers-based ethanol sensor with good reversibility and selectivity and fast response and recovery time. Electrospun LaOCl/NiO composite nanofibers have significant performance in ethanol sensing against CO, NO_2, H_2, NH_3, due to incorporation of NiO that catalyzes gas sensing reaction [42].

Many methods have been carried on to improve the sensitivity, response, and recovery time, for example, combining p-type with n-type metal oxide semiconductors to form p-n junction remarkably improving the sensing characteristic [43, 44], functionalizing the surface of nanofibers with catalytic nanoparticle (such as Ag, Pd, Pt) [45], and doping salts (KCl, LiCl, NaCl, and $MgCl_2$) into nanofibers, especially in humidity sensors [46].

3.2.2 Organic nanofibers

Organic polymers, especially conducting polymers (CPs) as an alternative to inorganic semiconductors, provide attractive features such as mechanical flexibility, easy processing, and adaptable electrical conductivity. Many research efforts have been dedicated to the development of nano-sensors based on CPs such as PANI, polythiophene, and their derivatives. However, CPs have poor solubility in common solvents, which restrict its application. Many routes have been developed to overcome this drawback, for example, incorporating CPs into other polymeric systems (such as PS, PEO, CA) or synthesized in other conducting forms (oxidized, reduced) [47]. The charge transport for CPs is primarily due to hopping mechanism. This hopping occurred because of changing polymer resistance in the presence of a sensing gas. This change can be due to chemical change (doping/de-doping), conformational change, or polymer swelling.

Electrospinning provides abundance of activated sites for CPs immobilization due to its unique features, such as large surface area, high porosity, and large stacking density. Pinto NJ et al. [48] demonstrated that the electrospun-isolated nanofibers of poly(3,4-ethylenedioxythiophene) doped with poly(styrene sulfonic acid) can be used to sense vapors (NH_3, HCL, NO_2, aliphatic alcohols).

3.2.3 Hybrid nanofibers

Hybrid nanofiber-based sensors have been developed to overcome the drawbacks of inorganic nanofiber-based sensors (require high-operating temperature) and organic nanofiber-based sensors (low sensitivity). Few researchers have investigated hybrid nanofibers-based sensing devices, and they got promising results in terms of sensitivity, response time, and reversibility [49, 50]. Researchers should devote their work to improve stability, selectivity, and reusability of the sensors.

3.3 Fiber optic sensors

All the aforesaid nanofibers-based sensors depend mainly on electrical sensing principle; however, in some cases, electricity is not suitable for the target sensing analytes. Hence, the importance of using optical sensors is necessary. Among the optical properties that have been utilized at sensors is reflectivity, refractive index, color, and absorption coefficient. Refractive index has been investigated by our work team, and it was effective.

Fiber optic sensor technology has been rapidly developed in the past 30 years due to the innovations in telecommunication, semiconductor, and electronics sectors that have significantly reduced the prices of optical components and stimulated the development of optical fiber sensor [51]. Optical fiber sensors are capable of measuring a wide variety of physical properties, such as chemical changes, strain, electric and magnetic fields, pressure, temperature, displacement (position), radiation, flow, liquid level, vibrations, and light intensity. Optical fiber sensors exhibit a number of advantages over the conventional electrical and electronic sensors:

- Are non-electrical devices
- Require small cable sizes and weight that enable small sensor sizes
- Allow access into inaccessible areas
- Permit remote sensing
- Immune to radio frequency and electromagnetic interference
- Do not contaminate their surroundings and are not subject to corrosion
- Provide high sensitivity, resolution, and dynamic range
- Offer sensitivity to multiple environmental parameters

It is believed that optical fiber sensors will replace the conventional devices for the measurement of various physical, chemical, and biological parameters. Optical fiber sensors are dielectric devices that are chemically inert. They do not require electric cables and are technically ideal for working in hostile media, and corrosive environment for remote sensing applications [52–55].

In the following section, we will display some of our team efforts of hybrid fiber-optic/nanofiber sensors developments.

3.3.1 Hybrid fiber-optic/nanofiber sensors

3.3.1.1 Bio-medical sensors

Petrík et al. [52] have produced SiO_2 nanofibers. The surface of SiO_2 nanofibers was functionalized with enzymes. **Figure 6** shows the SEM images of nanofibers with and without enzyme immobilization. The functionalized nanofibers were attached at the tip of *Y* junction plastic optical fiber. The setup of the experiment is displayed in **Figure 7**. One end of *Y* junction is attached to an LED as light source, and the other is connected to the photodetector. The light emitted by the LED is carried to the sensing point and partially reflected back. The reflected power was observed to be a function of enzyme concentration as shown in **Figure 8**. The detection limit was nearly 10% of the full initial level.

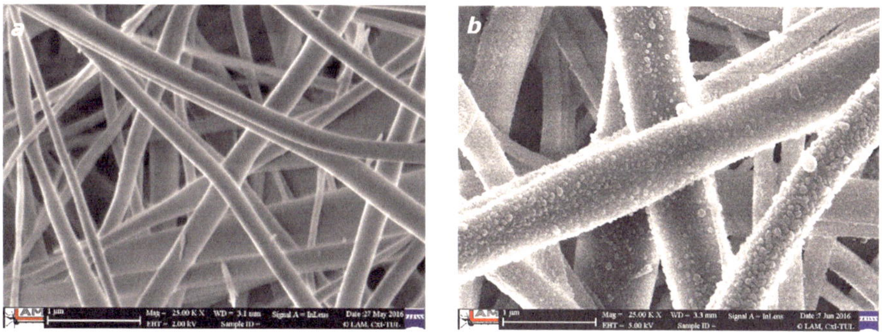

Figure 6.
SEM pictures of nanofibers without (a) and with (b) immobilized enzyme.

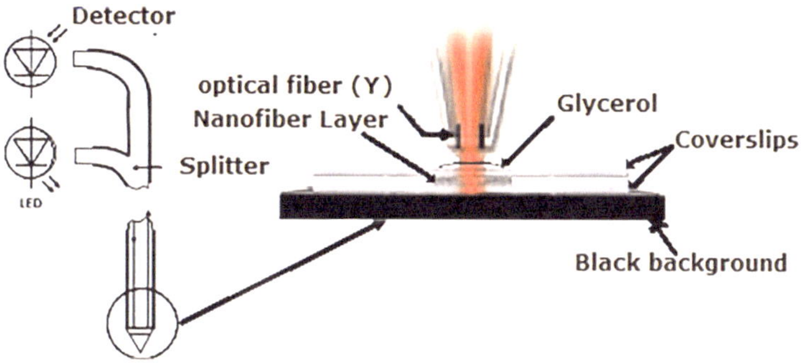

Figure 7.
Setup of the tested optical fiber sensor with an enlargement of the detection part.

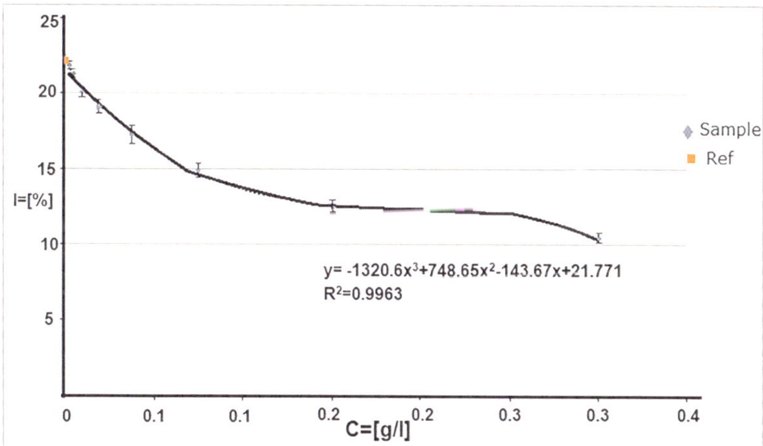

Figure 8.
Reflected intensity vs. concentration of a model enzyme-substrate.

The results of these experiments are very optimistic to the effectiveness of the prescribed optic fiber system with nanofibers. This system can be used as a basis of a wide family of optic fiber sensors sensitive to various chemical and biological substances. The proposed optical fiber sensor can be integrated into security systems for fast and cost effective.

Main advantages of the approach are as follows:

- Chemically inert materials—possibility to disinfect/sterilize
- Miniature dimensions
- Sufficiently high sensitivity

Another work effort from our group is a trial to estimate the water content in brake fluid using hybrid optical fiber/nanofiber as it will be explained later.

3.3.1.2 Waste-water cleaning process monitoring with nanofiber/fiber-optic sensors

We have used similar approach in a proof-of-concept study of using nanofiber/fiber-optic sensors for monitoring of waste-water bio-cleaning process.

Activity of bacteria in sludge water was monitored using online and offline optical fiber sensing system that utilizes the nanofibrous membranes. The used optics showed reasonable sensitivity levels to the slight changes in water compositions due to the presence of slurry matters and the formation of biofilms on the surface of the nanofibrous membranes. In general, the online setup showed better performance compared with the offline system that has inhomogeneous formation of bacterial films. As a future continuation of this work, other fibrous systems with higher compatibility and growing conditions for bacteria will be used. Also, functionalization of the fibers with elements that attract bacteria will be implemented in the upcoming work. Moreover, the experimental setup will adopt an online measurement system for mobilized and flowing bioreactors. The experimental setup is shown in **Figure 9**. An example of the sensor response to the bacteria activity is illustrated in **Figure 10**.

3.3.1.3 Silica nanofibers-based sensor concept for water content measurement

Focus of this research is to build and investigate an optical fiber sensor based on silica nanofibers prepared by a reliable and low-cost electrospinning technique to detect water content in DOT-4 brake fluid. To the best of our knowledge, this is a novelty study of optical fiber sensor to detect water content in an aqueous substance using electrospinning nanofibers.

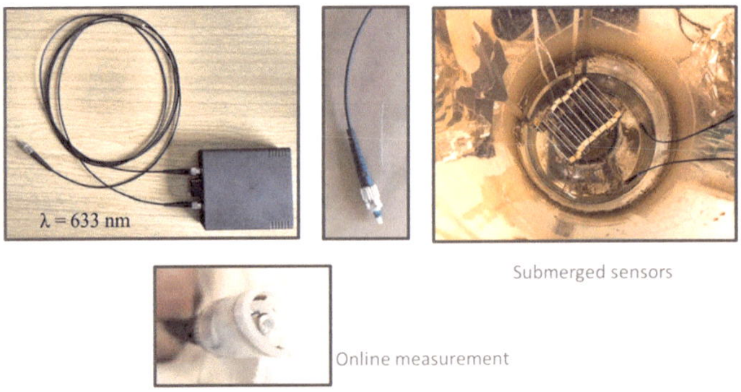

Figure 9.
Pictures for the used fiber optics and the online measurement setup.

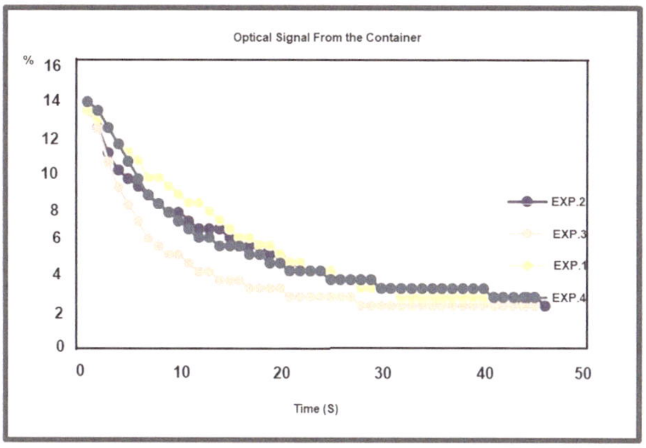

Figure 10.
Reflected intensity as a function of time detected by the hybrid fiber optic/nanofiber sensor.

The nanofiber processed by electrospinning has a larger specific surface area compared with conventional coating film, which can absorb a large number of water molecules. In some recently published articles [56, 57], dielectric properties of silica-based hybrid nanostructures and thin films have been investigated in which capacitance and dielectric constant act as a function of frequency. Batool et al. [58] studied the effect of RH on dielectric response of SiO_2 nanofibers; however, it is rarely investigated the effect of RH on refractive index of SiO_2 nanofibers.

The method used in this work involves utilization of silica nanofibers. The full description of preparation of the nanofibers can be found in patent WO 2017/186201 [59]. The composite PVP/SiO_2 nanofibers were left in the air for 24 h for hydrolysis of TEOS. Subsequently, PVP/SiO_2 nanofibers were annealed at 800°C for 6 h in furnace to obtain pure SiO_2 nanofibers. **Figure 11a** shows the scanning electron microscope (SEM) image of pure SiO_2 nanofibers after the removal of PVP, annealed at 800°C for 6 hours. The nanofibers have diameters ≈ 150- 200 nm. **Figure 11b** shows the energy-dispersive spectrum of SiO_2 nanofibers. The presence of atomic % of Si and O in the sample indicates formation of SiO_2 and complete removal of PVP. Si-O-Si bonds (siloxane groups) at 1087 and 797 Cm^{-1} become more intense after heat treatment as shown in **Figure 11c**.

Measurements were made with 0–7% water added to the brake fluid. The amount of water that was added to the brake fluid was determined according to dry basis moisture content (designated M_d in the text) is described by the percentage equivalent of the ratio of the weight of water (W_W) to the weight of the dry matter (W_d), herein is DOT-4.

Dry Basis Moisture Content is defined by Eq. (1):

$$Md = 100 \times (\text{Wet Weight–Dry Weight})/\text{Dry Weight}. \quad (1)$$

Commercial brake fluid tester was utilized for checking the percent of water content presented in brake fluid.

Silica nanofibers were glued on the tip of 2x Multimode optical fiber 50/125 μm, optical power meter as a source of input light, and a detector of the reflected light. OFS has been immersed in a brake fluid while changing its water content.

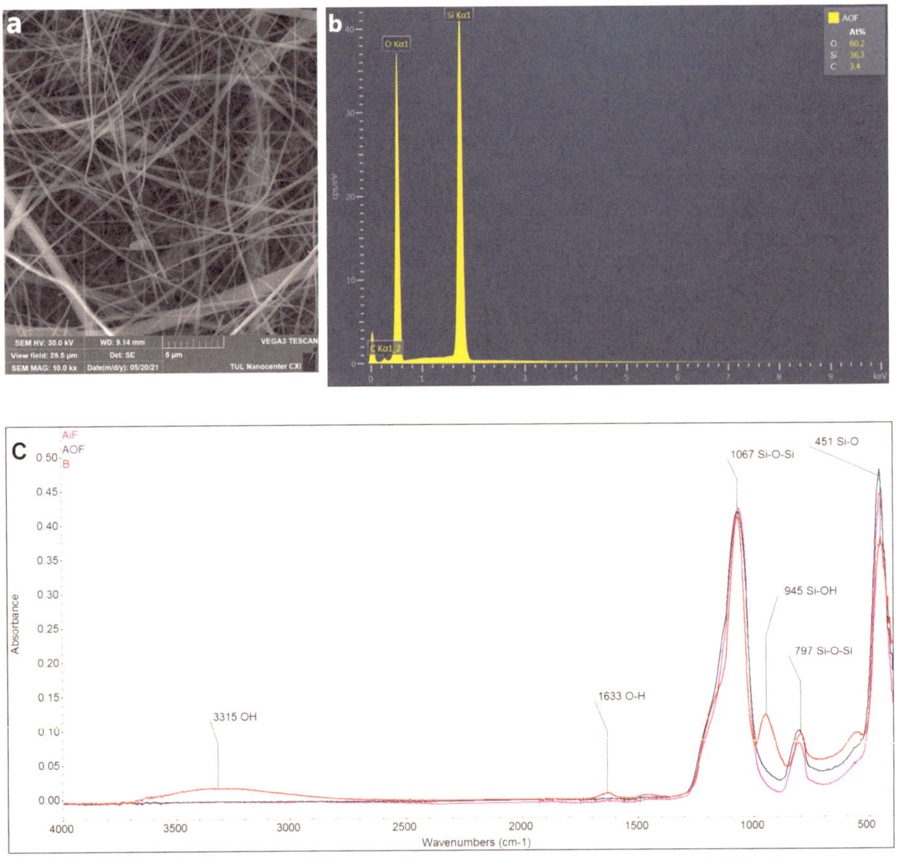

Figure 11.
(a) SEM images of SiO_2 nanofibers heat treated at 800°C for 6 h, (b) EDS spectrum, (c) Fourier-transformed infrared spectroscopy of TEOS/PVP electrospun fibers before (B) & after (A) heat treatment from both sides interior (i) & exterior (o).

Figure 12 shows the change in the power intensity as a function of water content in brake fluid which is almost linearly. That is probably related to water molecules that will be absorbed and concentrated in the pores of the silica nanofibers. This effect will alter the refractive index (RI) of the silica nanofibers, hence changing the optical power intensity. As a result, the accumulation of the water molecules will cause the increase of effective refractive index of the surrounding medium. This will lead to the leakage of the light through evanescent field [60, 61]. This proposed sensor based on reflected light intensity modulation. Based on Fresnel reflection, a proportion of lights are leaked when the sensor is in the liquid. This amount of light depends on the refractive index of the liquid. For normal incidence, the reflectance simplifies to the following equation Eq. (2).

$$R = \left|\frac{n_1 - n_2}{n_1 + n_2}\right|^2 \tag{2}$$

- n_1, n_2 are the RIs of optical fiber and exterior; n_1 are known by the fabrication of the optical fiber,
- n_2 is variable according to the exterior medium.

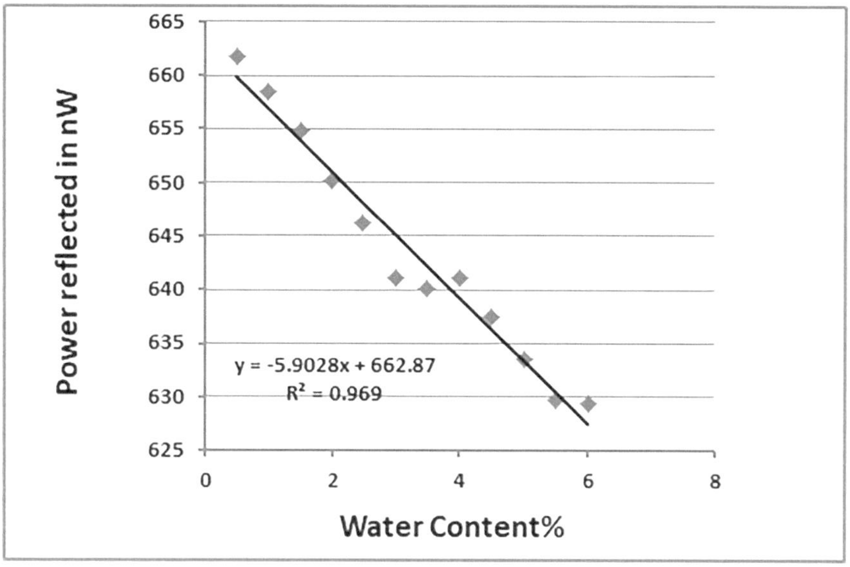

Figure 12.
OFS response to the water change in brake fluid in terms of changing light intensity.

The experiments are still going on while controlling humidity and temperature to assign the parameters, which could influence the accuracy and repeatability of the potential sensor.

4. Conclusion

Hundreds of papers are being published per year on "sensing" nanofibers.

Electrospinning looks like the most versatile method for their fabrication. Many unique sensor designs require just mm^2 of the nanofiber mat per unit (single use or multiple/continuous measurement). Definitely, they will not generate market opportunities for 1–2 m width production lines available on the market (Elmarco, Innovenso). The producers should probably consider the development of small volume special machines. Some of the "lab tools" the offer will probably fulfill the market needs.

Smart membranes/textiles are much more compatible with current production lines offered to the market. But the technological processes will be probably challenging and will need further development (chemistry, depositions of special substances, inter-operations, after-treatments, etc.).

Acknowledgements

This work was supported by the Ministry of Education, Youth and Sports in the Czech Republic under the "Inter Excellence – Action programme" within the framework of project "Micro-struCtural imaging as a Tool for modelinG fibrOus materiALS (μ-CT GOALS)" (registration number LTAUSA18135). Also, this work was supported by the Ministry of Education, Youth and Sports of the Czech Republic and the

European Union - European Structural and Investment Funds in the frames of Operational Programme Research, Development and Education - project Hybrid Materials for Hierarchical Structures (HyHi, Reg. No. CZ.02.1.01/0.0/0.0/16_019/0000843).

Author details

Stanislav Petrík* and Mayza Ibrahim
Technical University of Liberec, Liberec, Czech Republic

*Address all correspondence to: stanislav.petrik@tul.cz

References

[1] Anton F. Method and Apparatus for Spinning 2349950A

[2] Huang Z-M, Zhang Y-Z, Kotaki M, Ramakrishna S. A review on polymer nanofibers by electrospinning and their applications in nanocomposites. Composites Science and Technology. 2003;**63**(15):2223-2253

[3] Formhals A. Process and Apparatus for Preparing Artificial Threads. 1,975,504. 1935

[4] Formhals A. Method and Apparatus for Spinning, 2160962. 1939

[5] Anton F. Artificial Thread and Method of Producing Same, 2187306A. 1940

[6] Anton F. Artificial Thread and Method of Producing Same, 2187306A. 1943

[7] Norton CL. "Method of and Apparatus for Producing Fibrous or Filamentary Material," 2048651A. 1936

[8] Colleagues and Students. On the 100th anniversary of the birth of I.V. Petryanov-Sokolov Izvestiya. Atmospheric and Oceanic Physics. 2007; **43**(3):395

[9] Taylor GI. Disintegration of water drops in an electric field. Proceedings of the Royal Society of London. Series A, Mathematical and Physical Sciences. 1964;**280**(1382):383-397

[10] Taylor GI. The force exerted by an electric field on a long cylindrical conductor. Proceedings of the Royal Society of London. Series A, Mathematical and Physical Sciences. 1966;**291**(1425):145-158

[11] Taylor GI. Electrically driven jets. Proceedings of the Royal Society of London. Series A, Mathematical and Physical Sciences. 1969;**313**(1515): 453-475

[12] Simon EM. NIH Phase I Final Report: Fibrous Substrates for Cell Culture (R3RR03544A). Berlin, Germany: ResearchGate; 2017

[13] Merritt SR, Exner AA, Lee Z, von Recum HA. Electrospinning and imaging. Advanced Engineering Materials. 2012;**14**(5):B266-B278

[14] Wang XF, Ding B, Yu JY. Functional nanofibers in sensor applications. In: Functional Nanofibers and their Applications. Amsterdam, Netherlands: Elsevier; 2012. pp. 209-235

[15] Lin J, Ding B, Yu J. Direct fabrication of highly nanoporous polystyrene fibers via electrospinning. ACS Applied Materials & Interfaces. 2010;**2**(2):521-528

[16] Kessick R, Tepper G. Microscale polymeric helical structures produced by electrospinning. Applied Physics Letters. 2004;**84**(23):4807-4809

[17] Koombhongse S, Liu W, Reneker DH. Flat polymer ribbons and other shapes by electrospinning. Journal of Polymer Science Part B: Polymer Physics. 2001;**39**(21):2598-2606

[18] Jin Y, Yang D, Kang D, Jiang X. Fabrication of necklace-like structures via electrospinning. Langmuir. 2010; **26**(2):1186-1190

[19] Sun Z, Zussman E, Yarin AL, Wendorff JH, Greiner A. Compound core–shell polymer nanofibers by co-electrospinning. Advanced Materials. 2003;**15**(22):1929-1932

[20] Dror Y et al. One-step production of polymeric microtubes by co-electrospinning. Small. 2007;**3**(6): 1064-1073

[21] Zhao Y, Cao X, Jiang L. Bio-mimic multichannel microtubes by a facile method. Journal of the American Chemical Society. 2007;**129**(4):764-765

[22] Chen H, Wang N, Di J, Zhao Y, Song Y, Jiang L. Nanowire-in-microtube structured core/shell fibers via multifluidic coaxial electrospinning. Langmuir. 2010;**26**(13):11291-11296

[23] Kokubo H, Ding B, Naka T, Tsuchihira H, Shiratori S. Multi-core cable-like TiO_2 nanofibrous membranes for dye-sensitized solar cells. Nanotechnology. 2007;**18**(16):165604

[24] Mou F, Guan J, Shi W, Sun Z, Wang S. Oriented contraction: A facile nonequilibrium heat-treatment approach for fabrication of maghemite fiber-in-tube and tube-in-tube nanostructures. Langmuir. 2010;**26**(19): 15580-15585

[25] Yang X, Wang Y, Qing X. A flexible capacitive sensor based on the electrospun PVDF nanofiber membrane with carbon nanotubes. Sensors and Actuators A: Physical. 2019;**299**:111579

[26] Wang X, Li Y, Ding B. Electrospun nanofiber-based sensors. In: Electrospun Nanofibers for Energy and Environmental Applications. Berlin, Germany: Springer; 2014. pp. 267-297

[27] Yoon J-W, Choi J-K, Lee J-H. Design of a highly sensitive and selective C_2H_5OH sensor using p-type Co_3O_4 nanofibers. Sensors and Actuators B: Chemical. 2012;**161**(1):570-577

[28] Choi S-H, Hwang I-S, Lee J-H, Oh S-G, Kim I-D. Microstructural control and selective C_2H_5OH sensing properties of $Zn_2\ SnO_4$ nanofibers prepared by electrospinning. Chemical Communications. 2011;**47**(33):9315-9317

[29] Kanehata M, Ding B, Shiratori S. Nanoporous ultra-high specific surface inorganic fibres. Nanotechnology. 2007; **18**(31):315602

[30] Kim I-D, Rothschild A, Lee BH, Kim DY, Jo SM, Tuller HL. Ultrasensitive chemiresistors based on electrospun TiO_2 nanofibers. Nano Letters. 2006; **6**(9):2009-2013

[31] Zheng W et al. A highly sensitive and fast-responding sensor based on electrospun In_2O_3 nanofibers. Sensors and Actuators B: Chemical. 2009;**142**(1): 61-65

[32] Modafferi V et al. Highly sensitive ammonia resistive sensor based on electrospun V_2O_5 fibers. Sensors and Actuators B: Chemical. 2012;**163**(1): 61-68

[33] Song X, Wang Z, Liu Y, Wang C, Li L. A highly sensitive ethanol sensor based on mesoporous ZnO–SnO_2 nanofibers. Nanotechnology. 2009; **20**(7):75501

[34] Leng J, Xu X, Lv N, Fan H, Zhang T. Synthesis and gas-sensing characteristics of WO_3 nanofibers via electrospinning. Journal of Colloid and Interface Science. 2011;**356**(1):54-57

[35] Ruggieri F et al. Preparation of nitrogen doped TiO_2 nanofibers by near field electrospinning (NFES) technique for NO_2 sensing. Sensors and Actuators B: Chemical. 2013;**179**:107-113

[36] Wang W et al. Zinc oxide nanofiber gas sensors via electrospinning. Journal of the American Ceramic Society. 2008; **91**(11):3817-3819

[37] Das S, Kar S, Chaudhuri S. Optical properties of SnO_2 nanoparticles and nanorods synthesized by solvothermal process. Journal of Applied Physics. 2006;**99**(11):114303

[38] Liu XM, Wu SL, Chu PK, Zheng J, Li SL. Characteristics of nano Ti-doped SnO_2 powders prepared by sol–gel method. Materials Science and Engineering A. 2006;**426**(1–2):274-277

[39] Zhang Y, Li J, An G, He X. Highly porous SnO_2 fibers by electrospinning and oxygen plasma etching and its ethanol-sensing properties. Sensors and Actuators B: Chemical. 2010;**144**(1): 43-48

[40] Li Z et al. Highly sensitive and stable humidity nanosensors based on LiCl doped TiO_2 electrospun nanofibers. Journal of the American Chemical Society. 2008;**130**(15):5036-5037

[41] Fan H-T, Xu X-J, Ma X-K, Zhang T. Preparation of $LaFeO_3$ nanofibers by electrospinning for gas sensors with fast response and recovery. Nanotechnology. 2011;**22**(11):115502

[42] Hwang DK, Kim S, Lee J-H, Hwang I-S, Kim I-D. Phase evolution of perovskite $LaNiO_3$ nanofibers for supercapacitor application and p-type gas sensing properties of LaOCl–NiO composite nanofibers. Journal of Materials Chemistry. 2011;**21**(6): 1959-1965

[43] Wang Z et al. Improved hydrogen monitoring properties based on p-NiO/n-SnO_2 heterojunction composite nanofibers. Journal of Physical Chemistry C. 2010;**114**(13):6100-6105

[44] Xu L et al. NiO@ ZnO heterostructured nanotubes: Coelectrospinning fabrication, characterization, and highly enhanced gas sensing properties. Inorganic Chemistry. 2012;**51**(14):7733-7740

[45] Zhang H et al. Enhancement of hydrogen monitoring properties based on Pd–SnO2 composite nanofibers. Sensors and Actuators B: Chemical. 2010;**147**(1):111-115

[46] Wang W et al. Humidity sensor based on LiCl-doped ZnO electrospun nanofibers. Sensors and Actuators B: Chemical. 2009;**141**(2):404-409

[47] Wang S et al. Fabrication of electroactive oligoaniline functionalized poly (amic acid) nanofibers for application as an ammonia sensor. RSC Advances. 2013;**3**(12):4059-4065

[48] Pinto NJ, Rivera D, Melendez A, Ramos I, Lim JH, Johnson ATC. Electrical response of electrospun PEDOT-PSSA nanofibers to organic and inorganic gases. Sensors and Actuators B: Chemical. 2011;**156**(2):849-853

[49] Choi J, Park EJ, Park DW, Shim SE. MWCNT–OH adsorbed electrospun nylon 6,6 nanofibers chemiresistor and their application in low molecular weight alcohol vapours sensing. Synthetic Metals. 2010;**160**(23):2664-2669

[50] Wang Y et al. Ammonia gas sensor using Polypyrrole-coated TiO_2/ZnO nanofibers. Electroanalysis. Jun. 2009; **21**(12):1432-1438

[51] Yasin M, Arof H, Harun SW. Fiber Optic Sensors. Norderstedt, Germany: BoD–Books on Demand; 2012

[52] Lovětinská Šlamborová I, Kysela M, Petrík S, Vaněk T. Functionalized nanofibers as sensing Materials for fiber optic sensors, nanofibers, applications and related technologies—NART 2017. Conference Proceedings; September 25-27, 2017; Liberec, Czech Republic, Technical University of Liberec, 1. 2017.

pp. 109-116, 8 p [Online]. ISBN: 978-80-7494-393-5

[53] Turan JAPS. Optické vláknové senzory. 1. vyd ed. Petržalka, Slovakia: Bratislava: Alfa; 1991

[54] Udd E, Spillman WB Jr. Fiber Optic Sensors: An Introduction for Engineers and Scientists. Hoboken, New Jersey, United States: John Wiley & Sons; 2011

[55] Haus J. Optical Sensors: Basics and Applications. Wiley Online Library. Hoboken, New Jersey, United States; 2010

[56] Morales-Acosta MD, Quevedo-Lopez MA, Alshareef HN, Gnade BE, Ramírez-Bon R. Dielectric properties of PMMA-SiO_2 hybrid films. Materials Science Forum. 2010;**644**:25-28

[57] Wagner T et al. A high temperature capacitive humidity sensor based on mesoporous silica. Sensors. 2011;**11**(3):3135-3144

[58] Batool SS et al. Silica nanofibers based impedance type humidity detector prepared on glass substrate. Vacuum. 2013;**87**:1-6

[59] Ruzickova J, Tejkl M, Kudrova A, Vosatka Z, Buk J. Precursor fibers intended for preparation of silica fibers, method of manufacture thereof, method of modification thereof, use of silica fibers. Google Patents; WO 2017/186201 Al. 2016. p. 38

[60] Yuan T et al. Humidity sensor based on micro optical fiber array fabricated by electrospinning. Optics Communication. 2018;**427**:517-521

[61] Liu G et al. Low-loss prism-waveguide optical coupling for ultrahigh-Q low-index monolithic resonators. Optica. 2018;**5**(2):219-226